Fostering Institutional Development and Vital Change in Africa and Asia

Fred M. Hayward

Fostering Institutional Development and Vital Change in Africa and Asia

palgrave
macmillan

Fred M. Hayward
Washington, DC, USA

ISBN 978-3-031-04363-5 ISBN 978-3-031-04364-2 (eBook)
https://doi.org/10.1007/978-3-031-04364-2

This Palgrave Macmillan imprint is published by the registered company Springer Nature Switzerland AG
The registered company address is: Gewerbestrasse 11, 6330 Cham, Switzerland

Dedicated to:

My parents, Jean Port Hayward, teacher and mother who inspired us with creative projects throughout our childhood, and our father Herman Eliot Hayward, scientist, scholar, and the person who taught me how to write.

Acknowledgments

I am especially grateful to the people of Sierra Leone, Ghana, Madagascar, Afghanistan, Pakistan, and South Africa for their openness, thoughtful comments, and willingness to be observed over time. In none of these countries was I ever turned down for an interview or made to feel like an intruder by those with whom I worked. In Sierra Leone I want to thank: in particular, Michael Crowder, Banja Tejan Sie, Presidents Shaka Stevens and Joseph Momoh; in Ghana the Chancellor and Vice Chancellor of the University for Development Studies, and in Apiride, the Chief, Elders, and healers who let me into both the community and some of their secrets. In Madagascar I am particularly grateful to the then Minister of Education Haja Nirana Razafinjatovo, my colleagues in the Ministry of Education, and generous support from the government of Madagascar. In Pakistan I want to thank in particular Professors Atta-ur-Rahman and Sohail Naqvi. In Afghanistan I owe a special thanks to Deputy Minister M. O. Babury for his assistance and friendship over almost a decade of work there. In South Africa I am indebted to Professor Teboho Moja, Nico Cloete, then director of the Center for Educational Transformation, the Vice Chancellors of twelve historically black universities, Pirishaw Kamay, students who carefully tallied and retallied the votes, and my colleagues on the Electoral Commission.

Several people read drafts of this manuscript at various stages and made extremely useful suggestion which I have incorporated in this document. They include Professor Teboho Moja, Dr. Daniel J. Ncayiyana,

Marilyn Smith for her editing, Mujtaba Hedayet, Razia Karim, and former University California Regent Sue S. Johnson. I also want to thank the anonymous readers of this document for their useful suggestions and especially the thoughtful assistance of the publisher and staff of Palgrave Macmillan, especially Alina Yurova, Aishwarya Balachandar, Henry Rodgers, and Irene Barrios-Kezic.

I want to acknowledge the major funders of my research over the last fifty years. They include: The Ford Foundation, The Carnegie Corporation, The World Bank, the University of Massachusetts, Amherst, the Ministry of Higher Education in Madagascar, Midwestern Universities Consortium for International Activities, the University of Wisconsin Foundation, the Social Science Research Council Foreign Area Fellowship Program, several universities around the world, as well as USAID. Of course, none of these individuals or funders are responsible for anything written here—that rests solely with me. Finally, I want to thank my wife Linda Hunter, who put up with long periods in which I was away or seemed wedded to my computer, but she nonetheless remained a great supporter.

Introduction

Hope Fostered by Successes, Despair Among the Ruins of Failure

I have written about politics, higher education, and a variety of professional topics over the last fifty years;[1] and in these pieces I have often focused on leadership,[2] mostly national, and in retrospect, overlooked most of the personal and often life-saving stories of people at many levels who were critical to the change efforts as they did not seem appropriate to academic writing. There has been a tendency in much of the literature on development to focus on the elites and look at others involved only collectively rather than examining their contributions more deeply and exploring the diversity of the public and communities which help foster change and transformation.[3] Having thought about my research

[1] For example: Hayward, Fred M. (2020), *Transforming Higher Education in Asia and Africa: Strategic Planning and Policy,* SUNY, New York; and Hayward, Fred M., and Daniel Ncayiyana (2015), "Challenges of Graduate Education in Sub-Saharan Africa," *International Higher Education,* no. 79, Winter 2015.

[2] See especially my chapter 13 on leadership and transformation change in Hayward (2020) noted above.

[3] In his excellent book on political leadership, Robert Rotberg notes that "...greater attention should be paid to the critical role of leadership in the developing world." See Rotberg (2012), p. 2. I agree but want to go on to dissect the role of people at all levels, as well as communities, that make major change and transformation possible.

and writing over the last fifty years, I realized that this focus led to important omissions. In this piece, I have tried to rectify that by focusing both on my experiences with leaders and on people in general involved at every level of the change process. I have done this by going back to my notes over the last decades to explore both important leaders, and especially to focus on those in these societies at every level who have helped fostered critical change or create the conditions for change. In many respects that gives a more credible picture of these developing countries, the ideas, beliefs, and actions that affect change. In some cases, the lack of this support has also helped explain the failures.

My study is in the context of social theory, seeking to explore human actions and their relationship to development success, transformation, and failures. This effort does not belittle the importance on national or institutional leadership. Indeed, it builds on critical early studies in that area exemplified by works such as the classic on leadership by J. M. Burns (1978) *Leadership,* including the very useful work by Philip Altbach (2011) *Leadership for World Class Universities*, the excellent study by Robert Rotberg (2012) *Transformative Political Leadership*, and the important classic by Harvey Glickman (1992) *Political Leaders of Contemporary Africa South of the Sahara: A Biographical Dictionary.* I too have been unconsciously affected by the assumption that leadership is the key to development as in Hayward (2020). I have come to realize that the focus on leadership and elites leaves out critical aspects of the change process which is usually relegated to a brief mention of the importance of "followership" with few examples, if any. That has led me to suggest that development theory also needs to focus much more on the activity, successes, challenges, and risks undertaken by citizens, members of the communities involved, as well as local leaders.

The clearest examples of the importance of looking beyond top leadership in this book are the findings from South Africa and Afghanistan. Although Nelson Mandela is probably the greatest leader of this century, and I had the privilege of working with him during the elections that brought the ANC to victory, the successful defeat of the apartheid regime would not have happened without the actions of hundreds of leaders at the local level, along with people willing to put their lives on the line on behalf of change (and often losing them), and their solidarity and agreement on goals over decades. Similarly in Afghanistan, where the major transformations took place in higher education, and to some extent in health, it was the hard work of people at every level, the risks for, and attacks on, many of them, and the effort of two national presidents to keep politics out of higher education, that allowed transformation in higher education in gender equity, increased access, and most importantly quality improvement. I lay out the case for the crucial role of these actors in the chapters on Afghanistan, South Africa, and several other cases explored here. Thus, I would suggest that the theory of development of underdeveloped nations needs to go beyond leadership to look at both important leaders, and the broad role of the public disaggregated into its many parts.

My methodology derives from social theory, as noted above, and to some extent functionalism which has had me looking carefully at public actors and what they do to foster change in each of the six cases examined here, some of which failed in their efforts at change and transformation. This study has involved intensive reanalysis of the material I have collected in my research over the last fifty years as well as extensive reading and rereading of relevant research, additional interviews with some of those involved, and then rethinking the actions at multiple levels which fostered change. I have also explored how lack of such support has been part of what led to failure—along with other factors such as coups and external interventions. It is by making the distinction between leaders and public actors that one can explain how higher education in Afghanistan was transformed, including the admission of women, increased access, and updating quality—things that could not be done at that moment nationally even by the most successful national leaders. Some of this work also benefited from the theories of Amartya Sen (1999), *Development as Freedom*, which suggests that the idea of *freedom* should be linked to the development process arguing that the "freedoms of individuals" are the basic building blocks of development which is enhanced by the "effective use of participatory capabilities by the public." Thus, there is a critical need to focus both on the public at many levels as well as leaders where change, and transformation are sought.

In this manuscript I present information and recollections which I have recorded over the years both as a study of development and also as a biography of my experiences over fifty years of working in developing countries. The book is organized by country. My exploration begins with my first visit in Africa to Sierra Leone in the 1960s, then on to Ghana, Madagascar, South Africa, Pakistan, and Afghanistan. This analysis, vignettes, short essays, and stories illuminate the daily events in the developing world that are critical to both successes and in their absence often lead to failures. I also have explored the amazing access I had to several leaders which gave me unusual insights into their thinking and actions. Some of these assessments are in one sense insignificant viewed alone; yet taken together, they provide a picture of the conditions which make cultures livable, people lovable, and change possible. Some represent actions by expatriates, though most tell of the behaviors of leaders and local citizens and the impact of local events which created change, transformation, and a livable environment.

Unfortunately, a number of these occurrences did not result in solutions to pressing political problems, improve education, enhance the economy, or resolve major conflicts—though many did. Yet, as a whole, these pictures demonstrate ongoing public efforts in all these countries seeking to enhance these societies and improve daily lives.

Like the delicate red rose growing through the piles of rubble of a bombed-out building in Afghanistan, these events have remained in my mind over all these years and demonstrated the pervasiveness of the beauty, life, wisdom, and renewal that can be found even in the most dreadful of situations.

Fred M. Hayward

Contents

CHAPTER 1

Sierra Leone—First Impressions and Experiences in Africa

Sierra Leone in 1966 was the first African country I visited, thanks to a Ford Foundation grant for my dissertation. My focus was on political party competition in developing countries recently freed from colonialism—a comparison of Sierra Leone and Senegal.

I arrived full of romanticism about these newly independent states and eager to tell the story of these two countries—one a former British Colony and the other ex-French. Sierra Leone had a special relationship with Britain under colonialism with Freetown, Sierra Leone, a home for returned slaves, called Creoles who spoke Krio, a linguistically recognized language. They dominated politics under colonialism and maintained unique ties to Great Britain even after independence.

A Brief Overview of Sierra Leone History

Sierra Leone, on the West Coast of Africa between Liberia and Guinea, had contact with European traders from the mid-1400s. The Portuguese built a fort on an island there in 1882 to protect their trade. From that time on, substantial trade in slaves, fine wood, gold, and other metals was conducted with forts along the coast built by the Portuguese, Dutch, French, and British. The British eventually came to dominate that area. With the end of the slave trade by Britain in 1807, Sierra Leone was

F. M. Hayward, *Fostering Institutional Development and Vital Change in Africa and Asia*,
https://doi.org/10.1007/978-3-031-04364-2_1

conceived as a place for returned slaves by the British who provided funds to help establish Freetown, which was literally to be a free country under British protection. More than 50,000 freed slaves were sent to Sierra Leone from Britain, Nova Scotia, and slaves were intercepted from slavers on the high seas over the next decade. In 1897 the area around Freetown (often called the Peninsula) became a Protectorate and then part of a Colony of the British, with the area eventually becoming Sierra Leone, a Protectorate of Britain. Self-government did not last long after several attacks by local people who originally welcomed the returned slaves. As a result, authority over the area was given by the British to the Sierra Leone Company, a private for-profit business. Political, military, and legal matters took place with the help of British troops. The US sent returned slaves to neighboring Liberia which was similarly considered a free state under US protection.

Work in Sierra Leone

My work in Sierra Leone began a long career of studying Africa and working in ten African countries over the years. Flying into Sierra Leone on the day before Christmas was both an exciting and daunting experience. Since I had received a Ford Foundation scholarship, which was administered through the US Embassy in Sierra Leone, my wife and I were met at the airport by the Chief Cultural Officer from the US Embassy, Arnold Gordon, and a driver. Because of the hills around Freetown, the airport was built across Fourah Bay on flat land and thus involved a drive and then a ferry ride of half an hour across the bay to Freetown. We collected our baggage, changed the mandatory $200 to Leones (in theory to cut down on the Black Market), and drove to the Ferry.

Arnold Gordon was a great storyteller, and we talked the entire way about Sierra Leone and US politics, the US Embassy, growing up Creole—who were originally ex-slaves, pressure on Creoles for their domination, and the beauty of the beaches. The trip by ferry was stunning with Freetown on the other side of the bay looking small and pristine with houses ascending the hills. Fourah Bay College and the eight story Kennedy Building were perched on top of one of the hills—the latter a gift of the United States.

I think back to a trip to Sierra Leone nine months later aboard the Ferry with my two sons, aged 9 and 10, whom I had returned home to pick up in time for school in Sierra Leone. The boys were accustomed to Africans in our house, as I taught about and engaged in research focused on Africa at the University of Wisconsin, Madison. Students were invited over for a meal at least once a week. Nonetheless, to my surprise, my youngest child, Mark, came to me as we were standing on the deck enjoying the view and whispered in my ear. "Dad, I knew there would be Africans in Sierra Leone, but I did not think there would be so many!".

I was startled by his observation, yet looking around, I could see that there were merely five or six non-Africans on the Ferry with more than 100 Africans. This was an observation on my son's part—not a concern.

"You will probably be the lone non-Africans in the school for the children of employees at Fourah Bay College," I said.

"That's OK," said Mark taking in the whole scene as well as the incredible view from the ferry, "I was surprised that's all."

I was delighted that both boys seemed so happy with these new experiences. My older son, Kent, was amazed by the huge ant hills along the way to the ferry—all taller than he with some taller than I am at six feet two inches.

Now returning to the scene of our first arrival, Arnold Gordan and the driver took my wife and me to a two-story house which Fourah Bay College owned in the middle of Freetown and gave me the key to the second-floor apartment. The house was a lovely brick two-story building, recently painted white with great balconies at the front, a gas station on one side, and a school on the other. We were told that the Speaker of the House of Parliament, Banja Tejan-Sie, lived in an attractive multi-story, house across the street. He was a lawyer by training and the major figure in Parliament running the sessions and discussions. Arnold suggested that we go see him and introduce ourselves the next day. We followed his wise suggestion which ignited a friendship and introduction to Sierra Leone politics.

At that time, I viewed the newly independent states of Africa as harbingers of a new order that would reverse the damage of colonialism, bring quick economic development, and result in options for democratic governments that would be inclusive, open, and focused on all people. I thought these changes would be driven by a revulsion against colonialism, the freedom now offered, and encouraged by assistance from the US, the UK, and other entities such as the World Bank, the UN, and donors motivated by high expectations of new possibilities.

The housing manager kindly left breakfast cereal, canned sausages, non-spoil milk, margarine, orange marmalade in a bottle, and a loaf of bread. We were also given a note from Professor Michal Crowder, director of the African Studies Institute with which I was to be affiliated, saying he would pick us up in the early morning the day after Christmas.

For a moment, my wife and I felt lonely. We decided to go out on our second-floor front balcony to take in the view. The balcony was furnished with nice-looking wicker deck chairs. We overlooked Syke Street, which turned out to be great for people watching. We were no longer alone. Indeed, soon the President's motorcade passed by on his way home.

The date was December 24, 1966—the day before Christmas. A few minutes later, we heard a knock at the door. Our downstairs neighbors, the Marcus-Jones family came up to invite us for Christmas eve dinner. Marcus, as he was called, taught law at Fourah Bay College (FBC) and carried on a remarkably successful private practice; Ann ran a local charity.

We were delighted. Dinner with the Marcus-Jones family provided an enjoyable evening of good food, an overview of Freetown, and wonderful discussions with lots of laughter. Marcus received his Ph.D. in law at Harvard and was eager to find out about current happenings in the US politics. We also learned that he was one of the few lawyers who would take cases against the Sierra Leone police or government—and usually won. He told us about one of his favorite victories:

"I was asked to defend the diamond mining company against a suit by an employee who had burned his arm in an 'ordeal ceremony'[1] surprisingly used by this British company to identify whom, among the five employees in the secure diamond sorting and evaluation section, had stolen a number of diamonds. The drawers containing diamonds were weighted each day and in this case a number of diamonds were shown to be missing. That the man received severe burns on his arm was undisputed. The company suspected him but could not prove he was the thief—which he denied. The company thus decided to use a 'trial by ordeal,' which was customary in the region where the counting room was located, to find the culprit.

[1] There were many types of ordeals used in Sierra Leone, varying by location. Some involved putting the suspects arm in boiling oil, if burned the person was guilty. Others used a hot knife in the same way. Still others relied on ceremonies performed by a specialist often using spells they created, which were a kind of hypnosis usually getting the suspect to confess if guilty.

The counting room staff consisted of two Englishmen and three Sierra Leoneans. All five agreed to the trial by ordeal. The local specialist in such matters gathered the five of them outside where he built a fire and put a pot of oil on to heat. The oil became steaming hot. Each of the men was to put his arm in the oil. If he was burned, he was guilty.

They started off with the two Englishmen. Each man put an arm in, one by one, without getting burned. Then the first Sierra Leonean did the same without trouble as did the second Sierra Leonean. The final man put his arm in the oil and was severely burned. Thus, he was proclaimed the thief and fired!

The burnt man found guilty by ordeal sued the company for damages claiming he was unfairly hurt on the job and the company was remiss in using trial by ordeal. I argued, before the judge, that this was a voluntary effort, a traditional legal way to establish quilt or innocence, that all those involved freely agreed to the trail, that each went through the ordeal without problems except for this man. According to the usual traditional law of the area, that proved his guilt. I reminded the judge that under Sierra Leone law, traditional law was valid in court cases."

"What did the judge conclude?" we asked, curious about this method of justice.

Marcus-Jones replied, "The Judge said, 'I am not going to rule on whether this traditional law is valid or not, but I will rule against the complainant because, anyone foolish enough to stick his arm in boiling oil deserves to be burned. Case dismissed.'"

We all laughed at that careful rationale.

First thing on the morning of the 26th of December, Michael Crowder was at our house. We drove together up the steep narrow road to Fourah Bay College (FBC) for my first day of working with the African Studies Program. My wife went to the Archives to do some research. Michael Crowder was an Englishman who was a well-known African specialist. He had been recruited from Nigeria to Sierra Leone to set up the African Studies Program at Fourah Bay College. Michael was handsome, thin, and a man of strong views. He was noted for getting things done, as well as his outstanding writing about Africa.

December 1966

My First Forays into African Politics

Sierra Leone was off to a good start after independence from Britain in April 1961, with a multi-party democracy, relatively free elections, an active press, a well-functioning multi-party parliament, and an openness to foreign scholars. During the last years of British rule, Sierra Leone had a Parliament with members (MPs) from all over the country and a number of political parties. Independence was gained on April 27, 1961, with Dr. Milton Margai as the country's first prime minister, followed by his half-brother, Albert Margai, a lawyer and politician. Then Siaka Stevens was elected in 1967, removed in a coup organized by Albert Margai soon thereafter. That did not last long, and Stevens was returned to power. Joseph Momoh, former head of the military, became President by nomination of then President Siaka Stevens, who wanted to retire, on November 28, 1985, although Momoh campaigned as the sole candidate as a matter of principle. President Momoh worked quietly after his election to restore a multi-party democracy.

The first Prime Minister, Dr. Milton Margai, a medical doctor, worked hard to make the multi-party democracy work well. Unfortunately, Albert Margai, the next Prime Minister was not interested in democracy and established a one-party state (OPS), which was continued by President Siaka Stevens. He was followed by President Joseph Momo who worked to return the country to a multi-party democracy. Sierra Leoneans were proud of their progress during the first Margai regime, the economy was booming, and one could easily talk about politics and people's hopes for the future.

Sierra Leone enjoyed a special relationship with Britain during the colonial period. The people in Freetown, the town the British established, settled, and supported as a home for freed slaves, were given special status as quasi-British citizens. The citizens of Freetown spoke Krio—which developed over the years as a combination of English, Temne Yoruba, and words they created. It eventually became the lingua franca of Sierra Leone spoken by more than 80% of the population. Creoles dominated Freetown. The Freetown Peninsula was called the Colony and the interior the Protectorate, with much less British oversight. Eventually, the whole area became a British colony ruled from Freetown with a Governor and Legislative Council. Parliament opened upon independence in April 1961.

To start to understand Sierra Leone politics, I attended Parliament each day to listen to the debates. I sat in the balcony, which contained five rows of comfortable seats for guests. The Parliament building was round sitting proudly on a hill, built from beautiful local stone as an independence gift from the Israelis in their early efforts to gain recognition by African states. The inside walls were covered with local Ash wood which created a warm but formal appearance. A lounge was located at the back of the building with floor-to-ceiling windows looking out on the city and the huge bay, the largest in the world—big enough to hold the whole British Navy, called Fourah Bay.

The two major political parties in Sierra Leone at that time were the Sierra Leone Peoples Party (SLPP) which was the governing party, and the All Peoples Congress (APC), the major opposition party. In addition, several other small parties stood in opposition. Listening to the debates, I was particularly struck by speeches of Siaka Stevens, the charismatic leader of the opposition APC. I was eager to meet and interview him.

Since we had met our neighbor, the Speaker of the House of Parliament, Banja Tejan-Sie, a jurist, and well-known lawyer on our second day in Freetown, he offered to introduce me to Siaka Stevens and others. Later, the Speaker often gave me copies of government documents to read and we frequently talked politics. Banja Tejan-Sie was a distinguished-looking man, clearly in good physical condition, who loved the outdoors and was highly respected. Tejan-Sie owned a power boat and took us out on the bay on several occasions. When he saw me in the Parliament visitor's gallery balcony one day, he sent a clerk up with a note suggesting that I meet him in the Member of Parliament's (MPs) lounge at the break.

In spite of my confidence, I was nervous about the first interview with Siaka Stevens, realizing that if the conversation went badly, my research was going to be impossible. Although I had read a great deal about Sierra Leone before I arrived, I quickly realized the limit of my knowledge. Subtle power plays were taking place, and the SLPP was putting not-so-subtle pressures on the opposition parties, especially the APC, to join them in a one-party state (OPS). The ethnic politics of Sierra Leone—especially Creoles versus up-country indigenous people—seemed obvious to me until experience demonstrated otherwise. Sixteen different ethnic groups lived in Sierra Leone, the largest being Mende and Temne; among the others were the Konos, Vai, Limba, Sherbro, Fula, and, of course, the Creoles of Freetown.

I quickly learned that my understanding of ethnic tensions was simplistic. Siaka Stevens, like many up-country people, was sent by his family to Freetown at age seven to live with a Creole family to ensure that he would have a proper education. The Freetown Creoles made sure of sending him to the Albert Academy in exchange for a little housework on his part. Thus, he grew up in many respects, as he noted, part Creole and part Limba through his father with a Vai mother. He later married a Mende woman and often spoke of his multi-ethnic background. In addition to English, he spoke Mende, Temne, Krio, and Limba—all mutually unintelligible languages.[2] Stevens' mixed background was typical of many Sierra Leoneans.

As suggested, I met Speaker Tejan-Sie at the entrance to the MPs Lounge. The Lounge was full of MPs, advisors, and visitors packed around tables, the air heavy with cigarette smoke, with coffee cups and Star Beer mugs in hand on a typical hot, humid day, enjoying the air-conditioning and each other's company. Stevens was seated at a far table with two or three other APC MPs. He rose with hand outstretched to welcome both the Speaker and me, apparently open and ready to talk. The Speaker, who was widely respected by everyone, and I were greeted heartily. Banja Tejan-Sie introduced me as a young researcher who should be helped in any way Stevens could. Speaker Tejan-Sie then departed to his chambers.

I was struck by the fact that this group of APC MPs sat separately from other MPs in the lounge. Later, I learned that this was due to the growing pressure by the governing SLPP on the opposition APC to join the government. Other MPs were therefore fearful of being seen near any of the opposition MPs—let alone talking to them—lest they be accused of disloyalty to the SLPP government. A few APC members had already crossed the aisle to join the SLPP as a result of intimidation, bribes, and pressure growing out of the talk of creating a one-party state. The sole party if that occurred would, no doubt, be the governing SLPP. The new Prime Minister, Albert Margai, warned APC members, "If you do not join

[2] There are more than 2000 different languages in Africa, most unintelligible to each other from four different major language families (such as English and German being in the same family). According to a study by my wife, Prof. Linda Hunter, the average Sierra Leonean speaks 3.5 languages. There was little pressure for one of the Sierra Leone languages to become dominant because 80% of the population spoke Krio, which had become the *lingua franca*.

us now, you won't even get a bone."[3] Some members of the APC thought Stevens should bargain with the SLPP to join the SLPP government to secure as many positions of power in the Cabinet as possible. Stevens was opposed to merger with the SLPP and to the one-party state.

I had seen Stevens speak on the floor of Parliament several times before this meeting. He seemed relaxed, totally confident, and charming. He was often humorous, and full of stories and Krio proverbs. Referring to corrupt politicians he would say in Krio, "Wherever you tie a cow; there it will eat" ("*usai den tai kaw na de I de it.*"). Meaning: wherever you find a politician, there you will find corruption. Stevens was eager to talk about the pressure on APC members to "cross the aisle" and join the SLPP, the threat of the proposed one-party state (OPS) by the SLPP, harassment of APC members and the APC newspaper (*We Yone*, meaning *our own*) by the police—all part of the government push to force a one-party state "voluntarily." The SLPP's push for a one-party state was then felt to be based on their fear that they would lose the upcoming elections—they had become unpopular because of corruption under Prime Minister Albert Margai.

I was ushered to a seat by Stevens as he greeted me, "You are welcome to ask any questions you want. I have no secrets."

After expressing my appreciation, I asked him, "What is the status of negotiations with the SLPP on the one-party state?".

He laughed and replied, "It is nothing. The SLPP never should have raised the matter, for their own good. They should have gone about the business of organizing for elections and said nothing. The party made too much noise about the ops. We must hold an election soon. We are ready for that; we are strong."

The question seemed to energize him, so I followed up, "How will you explain your opposition to the one-party idea on the campaign trail?".

Stevens spread his arms wide responding in his formal West-African English with a slightly British accent laced with Krio, as if he were giving a speech in a rural village—or "up-country" as he would say—here in the lounge. "I will tell the people, now the big power of the central government is only for five years because it is so great. Thus, people merely have power five years at a time. When the five years are up, we get to consider how those in power have done—whether we agree with the way the SLPP

[3] Notes by the author circa February 2007.

ran the government or whether we want a change. If you don't like the way the SLPP has done things when you see the palm tree (the SLPP ballot symbol) say '*no thank you pa*' and vote for the rising sun (the APC ballot symbol)—vote for the APC. If there is a one-party state, you will not get that chance."

"How will you deal with the fact that the SLPP is using government power over chiefs to get them to control voting in their chieftaincies?" I asked intrigued.

"That is straightforward," Stevens said in his deep sonorous voice. The lounge was suddenly quiet with everyone listening to the APC leader. "We are not opposed to chiefdomcy. I will say, now the paramount-chiefs have power too. When the Englishman left, he left two keys. He gave one key to the people and one key to the chief. Now if you have a strongbox with money in it and you have two keys, you give one key to your wife and you keep one key. You don't give two keys to your wife." People listening to Stevens laughed heartily. "So, in Sierra Leone there is one key to the chief, and you keep one key. If you let the chief tell you how to vote, you have given him two keys."[4]

We continued to talk until the call came to resume the Parliamentary session. I was eager to pursue the topic at greater length, especially about his view of the role, prospects, and plans for the APC in the future, so I asked Stevens if we could meet again. He kindly invited me to his house the following day for a longer discussion.

Over the years Siaka Stevens and I developed a close relationship. Stevens soon discovered that I did not share with others the content of our discussions, and that I knew the landscape (including corruption and skullduggery). Thus, he did not try to hide things from me. He even showed me the deed to his house so I would know he paid for it and shared many APC documents, often sending the papers to my home by taxi. I would hear a knock on my door, and a taxi driver would hand me a box of papers, "The Pa wanted you to have these." Siaka Stevens maintained a special relationship with taxi drivers having defended them as a former union organizer and later Minister of Labour. Some of the documents he asked me to keep were his only copies—he wanted these papers kept safe since the police raided his house from time to time as part of

[4] Ibid.

their harassment efforts including destroying the records he maintained. He knew I would protect them.

This marked a difficult time for Siaka Stevens and the APC, with constant pressure on them by the SLPP government, even jailing of some APC members (mostly on trumped-up charges) and tempting others with offers of high office and other rewards if they were to join the governing party. Stevens himself was constantly threatened with jail. On one occasion, the taxi drivers overhearing plans to arrest him, surrounded his house with more than fifty of their taxis bringing hundreds of people with them forming a massive barrier and crowd around his house to deter the single truckload of police who came to arrest him. They quickly left.

Despite the SLPP threats, Stevens was a shrewd strategist and politician who was widely respected by people in all parties for his political skills, as well as those in the army and police—some of whom gave him advanced notice of actions against them ordered by the SLPP government. On another occasion, an early warning allowed him enough time to climb over a back wall and seek shelter, ironically, at the nearby house of a leader of the SLPP who opposed these pressure tactics against the APC.

Stevens presented himself as a human being concerned about the well-being of the people; a union man, out to protect the common people against corrupt elites. And he demonstrated that commitment time and again. That indeed was how most Sierra Leoneans saw him.

Stevens lived modestly, even when being president, with the exception of the house he built for himself late in his presidency. That house sat at the top of a hill overlooking Freetown, painted white, multi-story, and imposing—which must have cost a great deal of money (many said the construction was financed by Lebanese merchants). That was one secret I neither asked him about nor tried to unravel, though he denied the assertion to the press.

Siaka Stevens was often accused of corruption by his opponents. However, my impression was that any money he received from the Lebanese community (a major business group in Sierra Leone), other businesses, or others was used for his party—the APC. When he died, he was insolvent, having been sustained in his last few years by President Momoh and government grants.

During my stay, I met with Stevens frequently at his home, and sometimes at the APC office. The latter usually proved difficult since the office was always packed with dozens of people, offering their support, wanting help, and ready to protect the leader. Stevens was easy to talk to and willing to discuss almost anything. He spoke about how his career as a trade unionist led him to politics. "Trade Union work has made me what little I am today, and I have a duty to help the cause of the workers by all means possible." Stevens lamented about his perceived lack of impact when he was Minister of Labour during the British transitional government, "As Minister of Labour, I did not have the power to fix wage rates—I wish I had."

Over time, Speaker Banja Tejan-Sie confided in me. He thought that he and I were the only two people in Sierra Leone who could talk to anyone in any of the political parties because we were trusted and did not share our conversations with others. That perception was widely held and was to prove invaluable to me over the next decades of research in Sierra Leone.

During the 1967 elections, Siaka Stevens invited me to travel with him up-country which allowed us time to talk while driving in his car between campaign events. It also demonstrated how individual support allowed the APC to survive since, unlike the governing SLPP, they did not have access to government funds. Before we left a local gas station owner filled up the tank of all five of the vehicles (and a truck) with gas at no charge. That happened throughout the trip. Sevens wanted my advice on diamond mining, legally controlled by a foreign mining company (DeBeers), noting how much they paid to the Sierra Leone government every year, fearing they would leave if the amount were to be increased. We talked about how it was not the amount of money the company paid Sierra Leone that counted, but the degree to which the payment was a percentage of their profits—my guess was De Beers profited between 40 and 50%. Stevens had not understood that before and was delighted. He also explained his push for greater rural development, more schools, improved roads, health services, and workers' rights.

The days of campaigning were long and dusty bouncing along rural dirt roads, watching out for trucks coming the other way which often did not yield and were hard to pass on narrow roads surrounded by jungle or rice fields. Toward the end of the day when we were clearly tired, Stevens would try to recharge us by regaling the group with stories or reciting BBC news podcasts he memorized from a period during World War II when he worked briefly as a news broadcaster. The fact that his jokes and many of his stories were in Krio, had forced me to learn the language so that I was not the only one who did not laugh at his jokes. I borrowed a Peace Corps manual and quickly developed my fluency.

What this trip showed me was why Stevens was so sure of success in the forthcoming election. It was clear that the APC had tremendous broad public support, including many local chiefs, in spite of interference by the SLPP government. While Stevens was a remarkable, charismatic leader, he also had strong local leaders in the chiefdoms who mobilized people to come out and see Stevens when he campaigned, and that had a powerful multiplier effect since his speeches were so dynamic, involved give and take with the audience, and always left the crowd mobilized for the APC and recognized the corruption and self-interest of the governing SLPP. Stevens could not have done it without the depth and breadth of organization with the APC and his and the party's ability to mobilize those without strong political attachments, to vote for the APC—sometimes at some risks given the presence of SLPP thugs in some areas. Stevens and the APC also had a strong presence of women in its leadership and on the campaign trail. There was a tradition of strong women in Sierra Leone going back many years, especially given the power of "market women" who controlled much of the economy. Almost all of them were strong and vocal supporters of the APC.

During the election campaign, I also traveled throughout the country with SLPP Minister of Education, Salia Jusu-Shariff. The turnout for leaders of both parties throughout the country was huge and the public's level of knowledge about events occurring was amazing. While waiting for political leaders to arrive, village elders would organize music and dancing especially by women's groups, and sometimes performances by male Guinea traditional "ballet type" dancers who entertained us with incredible acrobatic leaps, flips, and complex gymnastics. Thus, we would arrive to a wonderful carnival atmosphere at many of the stops. Jusu-Shariff was a hardworking man of great integrity and honesty—some people complained that he didn't give out money like other SLPP leaders—and he responded that he only used his own money to campaign, something he was able to do as a successful businessman (bookstores and an accounting firm). My feeling after the elections was that if Jusu-Shariff had been head of the SLPP, it might have won.

Although from opposing parties, both Jusu-Shariff and Siaka Stevens took great care in laying out their cases for support. Indeed, it struck me that these rural people knew more current events than their Western counterparts—a conclusion I later confirmed in a systematic comparative survey of US and Sierra Leone citizens.[5] In interviews with people at several of the rallies I attended they expressed a clear understanding of the voting process, what their favorite party stood for, including some of the small parties they saw as basically individual candidate efforts to gain a seat in Parliament. They saw the election as important to their futures since government, in their view, played such an important part in national development. They were also cynical about the efforts of some candidates, not Stevens or Jusu-Shariff, to use bully tactics to scare them into support. In surveys done several years later, we were also struck by the fact that both women and men had strong opinions about political parties and government in general.

Not all the rallies were as orderly and open as those by Stevens and Jusu-Shariff which emphasized peaceful and open competition. Other rallies I attended included threats and attempts at intimidation by candidates in both major parties, such as, "If you do not support us, you will get nothing and some of you will lose jobs!" A few candidates traveled with menacing thugs, some of whom beat up opposition supporters.

We knew that the election results would be close. Charges of attempts to fix the results were raised especially in hotly contested areas. Nonetheless, the APC and several other opposition parties had trained poll watchers whom I saw on election day as I traveled around the country, carefully watching the proceedings, and calling out any attempted irregularities they saw. Their knowledge of the rules and their careful observations helped make sure the outcome was a correct reflection of the will of the people.

As the ballot counting neared completion, it was clear that Siaka Stevens and the APC had won the elections. I was leaving Sierra Leone for Senegal on the final day of counting and stopped by the APC office to say goodbye to Sevens. My timing was impeccable as he had just received a letter from Governor General Lightfoot Boston appointing him Prime Minister effective from that day, March 21, 1967, and inviting him to State House. Stevens happily showed me the letter amid an excited shouting, often weeping crowd of supporters. I headed to the airport finding the roads filled with throngs singing, sweating, and laughing in celebration of the APC victory. Stevens headed to State House followed

[5] Hayward, Fred M. (1976). "A Reassessment of Conventional Wisdom about the Informed Public," *The American Political Science Review*, July 1976, pp. 433–451.

by a large crowd and an escort of taxis as the first opposition leader in Sub-Saharan Africa elected head of state—a momentous event!

I traveled the hour-long flight to Dakar, Senegal elated that I had experienced an historic success for newly independent states. When I arrived in Dakar, however, I learned that Stevens almost immediately after being sworn in had been overthrown by the military in a coup planned with the former Prime Minister Albert Margai. I was devastated and disillusioned, not as an APC supporter, rather as a scholar having spent equal time with other candidates for office, including those in the governing SLPP, especially Minister Salia Jusu-Shariff. In spite of pressure and police harassment, an opposition party won a national election. That dream was rudely shattered by the military coup—a process that was to become increasingly common in the developing world to the detriment of the public.

Military rule in Sierra Leone was rife with difficulties, corruption, battles over authority among the officers, and disagreements among military leaders about holding new elections. Stevens and the Governor General were jailed, and later, ironically, Albert Margai, the former Prime Minister, was jailed as well.

Stevens was eventually allowed to go to England. After a year of military turmoil and disagreements, he was called back to Sierra Leone and reinstalled as Prime Minister (1967–1971). He continued to have overwhelming public support including of some in the defeated SLPP. Later he served as President from 1971 to 1985. The military returned to the barracks, to the relief of almost everyone. Among other things, the military had failed to gain public support, in part because of their internal bickering, in part because they didn't understand local politics, and in part because they were unable to govern effectively. Following my year in Senegal I returned to Sierra Leone for a few weeks to finish my research.

I immediately went to talk to then President Siaka Stevens to catch up. We met regularly, frequently talking for hours about politics, his past, and his hopes for the future. While not a highly educated man, he spent time at Oxford, read a great deal, and listened. He maintained a shrewd wisdom and ability to understand the minds of the public. Wherever he went, he drew big crowds. Especially importantly, he listened to them. Over the years, it struck me that being head of state was a lonely job in which a president could not confide in many people, if any. I provided an opportunity for that—a person he trusted. When he wrote his autobiography[6] in 1984, he asked me to write the introduction—to my surprise and delight—which I willingly did.

Over the following years, I went to Sierra Leone regularly. If I did not stop by Stevens' office within a day or two of arriving, he would send a motorcycle policeman to wherever I was staying to say, "the Pa wants to see you later today." Siaka knew everyone coming or leaving Sierra Leone. Of course, I would go, and we would always enjoy a remarkable conversation—sometimes with television present. His office was large, wood-paneled, with a carved desk behind which hung a large map of Sierra Leone showing potential oil deposits not yet verified—the President's hopes of becoming "Sheikh Shaki" as he joked.

When President Stevens heard that my 65-year-old mother was in Sierra Leone with a group of schoolteachers from California, he invited her to his office in State House for a chat, sending his limo to pick her up. They enjoyed a wonderful hour-long discussion which ended when Siaka, intrigued by Mom's small camera, asked her for it—which she gladly gave him. When I saw him six months later, he asked me to bring him some replacement batteries for the camera next time and sent a picture of himself for my mother. Stevens also met my two sons when he was taking his customary solitary week-end drive along Lumley beach in his convertible. We were swimming and waved to him. Stevens stopped and invited my family to visit him in State House—which we did to great fun, laughter, and photos with Grandma's camera. The fact that he drove himself alone from time to time is an indication of his popularity and sense of security.

Stevens was full of stories which he loved to tell. He said his strategy for integrating the country was to have a girlfriend from each ethnic group and a house in each region. He also joked about how he asked the police to arrest one MP for carrying a gun without a permit. He respected the man but felt he was growing too publicly critical. He laughed saying that he knew the MP had obtained a permit yet guessed he would not have the permit with him while traveling to his home in the far North in Kono. He would carry his gun between the seats for protection while driving through the lawless diamond areas. Stevens' hunch proved correct. The MP was arrested. So, the MP asked another MP's driver to go to Kono to

[6] Stevens, Siaka P. (1984), *Siaka Stevens: What Life Has Taught Me*, Kensal Press, Bucks England.

get his permit, which the driver did. However, instead of taking the gun permit to the MP under arrest, the driver brought the permit to Siaka Stevens who rewarded him and kept the document. The next day, President Stevens asked the police to bring the MP to him from jail, figuring he had suffered enough. Stevens gave him his permit, chastising him for not having the paperwork with him and for his overt criticism. Stevens then asked the MP to become Minister of Education—which the MP accepted knowing full well that this was another of Stevens' ploys.

Over the years Stevens would explore with me names of people he thought should replace him when he retired—the First Vice President, S. I. Koroma; then Jusu-Sheriff, a minister, then Joseph Momoh, head of the army. He never questions his right to pick the person to replace him as president in what was now a one-party state. When he actually decided to retire, he picked Joseph Momoh to replace him. The fact that President Stevens was one of the few heads of state who could actually retire in his own country was a mark of the widespread support and respect he continued to have.

On May 29, 1988, I was in Sierra Leone and on my way to visit my now retired friend Siaka Stevens at his Juba Hill home to write up some of the most interesting of his repertoire of Krio proverbs, an idea we had talked about doing for years. As I walked up to the door of his house, the police guards who usually waved me through, stopped me to say that the President was sleeping, and I should come back the next day. However, as I drove away, one of the senior officers chased me down in his car, knowing of my close relationship to the president. He reported: "The Pa done die." ("The President has died"). He went on to ask me not to tell anyone else until they could inform President Momoh who was up-country and unreachable at that moment. The news stunned me. Stevens seemed happy and active the last time I saw him. He was 83 and lived a full and successful life. I felt broken-hearted at his death—a charismatic leader, a real character, at times devious, yet for me a friend and mentor who wanted to make life better for the people of Sierra Leone. Over the years he had maintained broad support from individuals and groups throughout the country.

I attended his funeral which drew huge crowds from all over the country. People from all over were packed together for miles following his casket from the church service at a local Protestant church up the hill to Parliament where he was laid in state. A packed crowd filled the Parliamentary room containing his casket, including politicians, civic leaders, ambassadors, businesspeople, as well people from all walks of life—a fitting farewell for this colorful, people-oriented leader.

April 1992

The Demise of a Promising Government of President Joseph Momoh

The phone was ringing incessantly as I raced from my backyard in Washington, DC to answer before the ringing stopped. "Professor Hayward?" the voice at the other end quired.

"Yes, this is Fred Hayward," I responded, now recognizing the voice.

"Prof. Fred, this is Joseph Momoh calling you from Guinea."

"Greetings Mr. President. I am delighted to hear from you. I have been worried since the coup on 29 April 1992. I heard the Russians flew you by helicopter to Guinea, although I wasn't sure." I was flattered and honored that this important leader called me directly, especially under his difficult conditions.

He continued, "I called to thank you and admit that I should have listened to you. I ought to have taken the precautions you suggested. I could not believe that those officers I helped over the years, particularly Captain Valentine Strasser, who was leading the fight against the Revolutionary United Front (RUF) from Liberia, would do such a thing as overthrow a legitimately elected government—and now this coup. Thanks to you, we at least prepared an escape route out the back of the residence through the thick bush, difficult trails, and down the hill to the Russian Embassy. We made it there with the help of some of their troops. Then they flew me to Guinea where the President is taking good care of me."

"This is one occasion when I am sorry to be right Sir," I replied. "I am so happy to hear your voice and know you are safe."

My relationship to President Momoh, of Sierra Leone, began seven years earlier in 1991 when I received an unexpected phone call from Dr. A. K. Turay (AKT), assistant to the President and a Linguistics Department faculty member at Fourah Bay College (FBC). I met AKT when I taught Political Science there. He said that President Momoh wished to meet with me in New York City the next week when he would be attending a meeting of the United Nations. He wanted to discuss some confidential issues with me which AKT could not disclose over the phone. I said that of course I would meet him whenever he liked.

I arrived at the New York hotel at the appointed time. Once inside, a guard whisked me directly up to President Momoh's suite, passing the US Secret Service agents on site to guard him using the mirrors in the hallways to see who was coming and going.

The President was dressed in a dark blue suit, looked relaxed, delighted to be in New York, and pleased to see me. He was a handsome man, and even his suit could not hide his muscular body. I smelled the remaining scent of his Jollof Rice lunch (chicken, peppers, rice, tomatoes, onions), as I entered which made me feel I was back in Sierra Leone.

"Greetings Prof. Fred, I am delighted to meet you. I have heard and read a great deal about you. Have a seat. I want to talk with you about rather significant issues."

"I am honored to finally meet you and look forward to talking about these issues. How can I be of assistance?"

"Let me start with a little background. As you know, as head of the Armed Forces I was often asked to undertake a number of tasks that were difficult for me—especially during the Presidency of Albert Margai, who used both the police and army to harass opposition political parties before he established a one-party state. I hated the use of violence and did everything in my power to limit it while maintaining the trust of President Albert Margai under the One-Party State. I worked with my troops on ways to control crowds and demonstrators with wicker shields to deflect stones and push back mobs. Only officers were armed, and we used tear gas solely for extreme cases. Even so, some people were hurt, and the police killed several demonstrators, shooting without authority. That sickened me and made me promise that If I were ever in a position to end that violence and intimidation, I would. I also promised myself that I would try to restore the multi-party system if I ever had the opportunity."

"That brings me to the reason I have asked you to meet. You know a great deal about African politics and Sierra Leone and have written about elections and democracy.[7] I want you to help me and a small group of advisors figure out how to move Sierra Leone back to a multi-party state—with an active Parliament and individual rights including freedom

[7] For example, Hayward ed., *Elections* in *Independent Africa* (1987) and multiple articles.

of speech and the press, as well as civil liberties. Would you be willing to do that?"

"Yes, I would be most pleased to assist with that effort. I have several short consulting commitments ahead yet think I could meet those obligations and spend most of the year with you. This is good timing since my wife is working in Wisconsin and our children are off to college. Returning to Sierra Leone where I taught at FBC and wrote my dissertation looking at the post-colonial governments in Sierra Leone and Senegal, will be a treat."

"I am delighted. Thank you, Dr. Hayward!" said President Momoh. "No one other than AKT knows about this initiative. I don't want word about this project to spread until we have all our plans ironed out."

Arrangements had been made for my flight, and in short order I was landing in Sierra Leone via London. I was met at the airport by my old friend from dissertation days, Abdul Karim Turay (AKT), now advisor to the President as well as continuing his work as Professor of Linguistics at the College. I was taken to the Presidential guest house to drop my things and then to AKT's home nearby for one of his wives, Jasmin's, delicious lunches of rice, Barracuda, tomato sauce, onions, and hot peppers. She asked me if I wanted the head of the Barracuda—regarded there as the greatest delicacy, especially the eyes. I said, "No thank you, I will leave that for you. You deserve it." The whole house smelled of the marvelous meal which I am sure was in preparation from early morning. Coupled with the warmth and hospitality of AKT and Jasmine, I was thrilled to return to Freetown!

Abdul Karim Turay (AKT) was a critical part of efforts to move to a democratic multi-party state. In his quiet way, he had influenced President Momoh and encouraged his democratic instincts. President Momoh's desire to reestablish a democratic government was surprising in many ways. He was a lifelong military man, had been head of the armed forces for many years, and had gone to a military academy in England. But rather than turn him in to an authoritarian leader, his experiences had convinced him, as noted above, that democracy was essential to national development, and he was committed to establish it. It was AKT who helped put together a strategy to get that done, ensure careful preparation before the proposed changes and referendum on the issues were announced. AKT was a popular, thoughtful faculty member at Fourah Bay College (FBC) who had come to know the President when he was head of the Army when giving lectures to the officers during their ongoing training.

They built up a friendship of many years before Major General Momoh became president, one of mutual respect and trust. AKT was also honest and refused to take money from those who came to bribe him. I worked in his office for more than a year and if someone brought money, as several Lebanese in particular did (even suitcases full), AKT would have his military bodyguard take the money and put it on the refrigerator in the next room. By the end of the day the money would be gone because AKT would give it to people in serious trouble who came to him for help during the day, including students who lacked funding for food, books, or lodging. Once again, he never touched the money. The hand-outs were done by his bodyguard. He regarded this as his Robin Hood effort.

AKT had gained weight since I last saw him, and he told me, "I am trying to get as big as Jasmine who is 8 months pregnant." He was indeed close to equaling her at this point. During the meal he said, "We are going to discuss whether or not the President should form a new political party because of the APC's history of corruption.[8] President Momoh wants you to prepare a balanced list of pros and cons before the meeting tonight."

Going back to the guest house after this delicious lunch, I prepared a list of ten reasons to form a new political party and ten to stay with the APC trying to make each list equally appealing. The meeting that evening, with a small advisory group on democracy, moved along and the President was pleased with the discussion. Given the corruption of the APC, forming a new party was tempting. However, given the existing successful APC party apparatus all over the country with broad support and clear representatives he could see a clear advantage to sticking with the APC and trying to reform it. In the end, President Momoh chose that option, especially focusing on building on the public support the APC had in much of the country.

Campaigning for President when he was the only candidate had endeared him to many people and garnered increased support for the APC. He immediately stopped use of the military and police to intimidate people for political gain, which made him especially popular—something he could do well as former head of the national army. Suddenly there were a dozen new independent newspapers.

[8] Under Siaka Stevens, many of the senior members of the APC had become quite corrupt and Stevens did little to stop them unless he felt they were too ostentatious about it.

I worked with the President and AKT on an almost daily basis in State House (the Presidential Office) on rules for the elections, new laws to allow political competition, freedom of the press and speech. We worked quietly laying the groundwork by freeing the press right away and sending home the few political prisoners incarcerated in the Pademba Road Prison. I also drafted a few speeches for the President to give in Parliament and on the radio. In one which was broadcast from State House with all the Ambassadors and other dignitaries present, I included the noble lines from Abraham Lincoln: "Government of the people, by the people, and for the people." We were amazed to hear the thousand plus people gathered outside State House to listen to the speech, chant the phrase which they knew by heart, in unison. All the dignitaries and ambassadors were as amazed by this as I was. The public response showed that people were eager to return to democracy. We then worked on a new constitution with a separate committee set up for that purpose, procedures for an electoral commission, conditions for the registration of political parties, and then candidates for Parliament. My year was incredibly busy and rewarding. The year saw increased support for the Government and the President.

President Momo had set up a small committee of people working in various roles in Sierra Leone, finance, law, an MP, several senior administrators, and me. We met weekly, or more often, in one of the President's guest houses, with no formal announcement, and in great secrecy. President Momoh felt that if information got out about the plan to reintroduce multi-party democracy, people in his own party (the APC) and others would do their best to scuttle it. He wanted to have all the plans in place before announcing the proposed changes in public. The committee worked hard with evening meetings often continuing until close to midnight. There was debate on how the organize the process of change. It was finally decided that the first effort after the announcement would be a referendum on a new constitution. If that succeeded, the President would move on to allowing additional political parties to form. At that point, the APC was the only legal party.

One thing that struck me during these meetings was the President's strong commitment to democracy, as was the case of all members of the committee. Thus, the decision to hold a national referendum on the new constitution which included legalization of political parties which met reasonable requirements, freedom of the press and speech, and other guarantees of democracy. These things could have been put in place by

Presidential decree, but the President wanted none of that. Even our committee was an important way to bring many different voices into the process, involving a great deal of debate, sharp difference of opinions on issues such as whether to outlaw the former governing party, the SLPP, for its corruption anti-democratic legislation, and mismanagement—the committee did not suggest he do that. Though the President had the final word, I was struck by how much this was a collective process, with all of us able to express our opinions, debate them with other members, and reach a consensus before the President made his commitment one way or the other. In each case it represented a consensus view.

I think that the care taken to prepare for multi-party democratic government was part of its success. The constitution passed in the national vote overwhelmingly. Seven political parties, including the existing APC, a reestablished SLPP, and five new parties, were set up once they became legal in early 1992. There was a great deal of excitement among the people about the return of elections, free speech, and freedom of the press. The dozen or more newspapers set up after the constitution passed were often acidic and critical in their analysis with everyone taken to task at some point. Those involved were relieved and delighted with the progress being made.

Unfortunately, that joy was moderated somewhat by the consequences of an attack on March 23, 1991, by a rebel group from Liberia led by Charles Taylor's National Patriotic Front of Liberia (NPFL), who wanted Sierra Leone gold and diamond to finance their war against the government of Liberia. By April 1992, about one-third of Sierra Leone was under their control or a no-man's land. The Sierra Leone Army proved weak in their defense of the homeland, even with troops and air support from Ghana, Guinea, and Nigeria. The NPFL troops were especially brutal, kidnaping your men to be soldiers, young women as sex slaves, and killing many villagers in the areas they controlled. They were especially noted for cutting off the hands, arms, or legs of those who resisted as examples to others not to resist them. That legacy of brutality, not to mention the loss of limbs, lived on for years after the war in the damage to these young people's lives, many now orphans, some with mental illnesses, some of whom had been forced to commit terrible atrocities. While by late in 1992, the war was at a stalemate, it greatly affected the government's efforts at democratization and posed a serious question about whether or not elections could be held in all of the country. President Momoh worried about even having the elections as then planned since most of the areas under NPFL control were SLPP strong-holds, and he feared if he called Parliamentary elections he would be seen as acting in a partisan way in favor of the APC and against the SLPP. He was torn by this dilemma as were most of his advisors.

On a Sunday while working in the President's guest house, I heard a heavy knock on the door. To my surprise, the visitor was the President's secretary. He was pale, clearly upset, and sweating profusely—not his usual jovial self. He had never come to see me before. I realized there must be a crisis for him to drive from downtown up the hill to the guest house. I invited him in immediately. "What can I do for you," I asked.

"I received a call from an ambassador. Their intelligence has learned that Captain Valentine Strasser and several other officers are planning a coup. I know that the President will not believe me if I tell him this news; but he will believe you. You can call the ambassador to verify the intelligence. I told him you would call. Here is his phone number."

I called the Ambassador who confirmed the intelligence and added more details about how the insurgents were moving armored vehicles from the North where they had been fighting the rebels from Liberia, toward the capital Freetown, in the guise of a maneuver related to that struggle. They were intending to overthrow the President.

I immediately called President Momoh on his private line, "I need to talk to you urgently even though it is Sunday. I must come to your house right now."

I arrived and was hustled upstairs to his study with a desk piled high with documents. He had set up a card table for the two of us. I could smell the fresh coffee we both enjoyed in two steaming cups. He stood up to greet me and ushered me to a seat. "What can I do for you?" he asked, curiously.

"I have come to warn you, on good information, that a coup has been planned in the next few days. The coup is being led by Captain Strasser and several other officers. They are using armored cars which are now headed toward Freetown from the northern battlefield."

"I can't believe that Captain Strasser would be involved in such a plot. I saved his life by sending my helicopter to pick him up at the front lines in the Eastern Province when he was wounded. I am sure he is loyal to me."

"Well sir, frankly, when I arrived here just now, I noticed only three soldiers on guard instead of the usual twenty, and the tank normally in the driveway is gone," I added, "The ambassador seemed certain of his information. At least, you should ask for some Nigeria troops here to help against the invaders from Liberia, to supplement the few guards who are here. I know the Nigerians have at least one hundred soldiers at their embassy not far away."

"I will," he said in a mournful tone. "Thank you for coming Prof. Fred and for all your help. I know you are headed on leave in a day or two, so I look forward to seeing you back here in about two weeks."

In the meantime, his assistant AKT went to talk to friends in the military garrison not far away and was warned to leave quickly because a coup was underway. He too then warned the President. The chief of police organized the police to oppose the military force but before he could do so was killed by local military leaders.

Two days later, after I left on my leave, on April 29, 1992, the coup was successful, and the President was overthrown. He managed to escape with Nigerian assistance to adjacent Guinea, where he maintained a long-term warm relationship with the Guinean President, as he later explained on the phone. This led to a long period of military coups in Sierra Leone, countercoups, and rampant corruption—all of which devastated the economy as well as the political system. It also broke down the strong system of public support built by the APC and to some extent the SLPP, given the rapaciousness of the military and its constant internal upheavals. Eventually elections were held once again, with the APC winning. But by this point the public was fed up with Government and those who won were not very interested in democracy or reform but in their own well-being and gaining wealth in any way they could. The public grew even more cynical and distrustful of Government—with good reason.

Unfortunately, Sierra Leone was also soon beset by the Ebola epidemic which killed more than 4000 people[9] from May 2014 to early 2016. As a result, the economy deteriorated by about one-third. However, in 2019 the economy grew by 4.2% thanks to recovery in the mining sector and in agricultural exports. The SLPP won the election in April 2018, with the APC and SLPP each sharing about half the members of Parliament.[10] At the time of writing, politics remained unstable, and corruption was not

[9] World Bank. (2018), "Country Report," p. 4, downloaded April 11, 2020, from: https://www-bti-project.org/en/reports/country-reports/detail/itc/sie/itr/wca/.

[10] World Bank. (2020), "The World Bank in Sierra Leone," downloaded on April 11, 2020, from: https://www.worldbank.org/en/country/sierraleone/overview.

under control—although several review commissions were underway to assess the problems.

While the situation in 2021 was slightly improved, the political conditions remained precarious with both major parties in disarray and several new parties with small numbers of members of Parliament involved in the fray. As the World Bank noted in its report in 2019, "The country continues to face daunting challenges of enhancing transparency in managing its natural resources and creating fiscal space for development. Problems of poor infrastructure and widespread rural and urban impoverishment persist despite remarkable strides and reforms." For the first time in many years there was some hope for both democratic progress and economic improvement with growth in 2019,[11] and with the recently elected SLPP government striving to deal with corruption. Nonetheless, the public had become so disillusioned and cynical that no one had much public support and anarchy reined in much of the country. Hope has not returned to the country or to ordinary citizens as I write this. Added to that has been the scourge of Covid, for which the Sierra Leone health services were woefully unprepared. That has only made public anger and cynicism worse with growth in inflation of 17.1% and growth of GDP falling by 4% in 2020 though it is expected to rebound by about 4%. It also suffers from high youth unemployment.[12]

Unlike most of the other countries examined in the following pages, there were no longer any active citizen groups or major critics leading the way to reform. In many respects this is an excellent example of what happens when anarchy reigns and no individuals or groups step forward to demand or bring about order and improvement of the sort we will examine in later chapters on South Africa, Ghana, and Afghanistan. It also demonstrates the long-term destabilizing effect of the use of the gun—military coups.

Thinking back to my time in Sierra Leone I am often amazed that I was able to maintain such a close relationship with Siaka Stevens. I was 27 years old when we met, working on my Ph.D. In later years, I established a similar close relationship with a number of other politicians in

[11] Ibid.

[12] World Bank. (Oct 2021), "The World Bank in Sierra Leone," World Bank, Washington, DC, pp. 1–3.

Sierra Leone and in several other countries, as well as with several Ministers of Higher Education. I believe these relationships blossomed because I was open, accepted people as they were, did not try to change their views or talk to others about what I learned. I also did not write about my experiences until much later and was friendly and respectful to all of them regardless of their ideology or points of view.

Looking back, I am also struck by the efforts of both Siaka Stevens and Joseph Momoh to improve the economic position of the country and their ability to mobilize people in general. President Momoh was especially committed to returning the country to a multi-party democracy and had significant success in moving in that direction. I am saddened by the limited successes of Stevens, and the destruction of President Momoh's efforts to return the country to a multi-party democracy. I also remain amazed by the close relationship I was able to develop with both Presidents and the role I was able to play in the latter case in the efforts to return to effective democracy. Somehow, an excellent chemistry and trust developed between us that is hard even today for me to explain. That closeness remained even after both people were no longer in power. Part of it grew out of the respect I developed for both presidents as I got to know them and the fact that their goals, and my hopes for the future of Sierra Leone, were the same. Both men were able to live in Sierra Leone in later years without fear and remained popular—a rare feat in much of the underdeveloped world.

Overall, however, Sierra Leone, which in its early days was a beacon of democracy and public participation, had managed to turn the public against government in any form and operated with a lack of broad participation and a national commitment to democracy. Part of Stevens' success was his charisma and ability to sense what people were thinking in various parts of the country and act accordingly. He also benefited from broad public support and leaders active on behalf of the APC throughout the country. President Momoh too worked hard to develop broad participation including an overwhelming vote of support for the new democratic constitution, but was hindered by the corruption in the APC which he only began to eliminate, and the lack of other leaders at various levels committed to an open, fair, and effective democracy. There were a few public advocates for democracy from faculty members at Fourah Bay College and in the press, but that did not mobilize sufficient public support or latter a commitment to a government rife with corruption.

Add to that the various coups, which only further created general alienation from government, and the country was unable to return to its previously successful democratic state. As such, Sierra Leone continues to suffer. There has been recent work of a White Paper on education (the Kwame Report), but at the time of writing, I am unable to get a copy of it and so far, no actions seem to have been taken as a result of it. This case demonstrates how loss of public trust and support, even with outstanding leadership, can undo previous democratic successes, and be much harder to rebuild than to lose. Elections in 2018, with an SLPP victory but a strong APC showing and World Bank supported efforts to improve both the economy and public life, may bring some positive changes.

CHAPTER 2

Ghana, a Model of Development

The Carnegie Corporation had asked me in 2003 to help several Ghanaian Universities with strategic planning, starting with University for Development Studies (UDS). I had worked in Ghana earlier, in 1992, to help prepare for, and supervise national elections, supported by the US International Foundation for Electoral Systems. I also taught at the University of Ghana as a Fulbright Professor in 1970–1971.

Arriving at the airport in Accra, Ghana, I was met by the driver of the Vice Chancellor, F.B.K. Kaburise (VC) of the University for Development Studies (UDS) who held a sign reading "Prof. Hayward." I greeted him happily wondering how he managed to get through security, then I saw he wore the VC's VIP tag.

The driver introduced himself saying: "I am Kofi Saffu, I am here to help you with the baggage and then we will drive to the University in the North together, probably staying half-way there overnight in Kumasi." Kofi took my briefcase as is expected in Ghana, and we headed to claim the luggage and go through customs. We waited and waited—I had that sinking feeling in my stomach watching the luggage come off the carousel—and indeed, no luggage arrived. We went to the Ghana Airways

F. M. Hayward, *Fostering Institutional Development and Vital Change in Africa and Asia*,
https://doi.org/10.1007/978-3-031-04364-2_2

counter to report the missing baggage. The staff person politely explained that they had been looking for me as British Airways sent a message that my baggage was in London and would be sent on the next flight. This was more of a problem than usual since we needed to drive 900 miles to UDC. The next flight from London would arrive in two days. The trip up and back to UDC was at a minimum one day each way depending on the speed of the driving.

I asked Kofi, “Can I use your phone to call Vice Chancellor Kaburise and explain the problem. My bag has all of my clothes and most of my diabetes medications, although I always carry enough for two days in my briefcase.”

“Let me call him for you,” Kofi offered. “He is expecting me to call to say I have picked you up. Here he is on the line,” and he handed me the phone.

Vice Chancellor Kaburise greeted me with a cheery, “Welcome to Ghana. We have been looking forward to having you here. We are a bit stuck on how to write the strategic plan which Carnegie requires for funding.”

“Vice Chancellor,” I said, “I am here to help you. The short-term problem is that my bag was not on the flight and is coming in two days on a British Airways flight. I could come without my belongings, but according to the airline officials I am the only one who can pick up my baggage with my passport and then pass through customs. For me to come to UDC and then have to make a two-day trip back and forth to Accra to get my luggage doesn’t seem to make sense. Shall I send your driver back and remain here until my bags arrive?”

“I am sorry to hear that,” said the VC, “more for you than for us, having encountered a similar problem myself recently. Why don’t you stay there? I have other drivers here, and the driver can help you pick up a few urgent things you need while in Accra.”

“That sounds good,” I replied. “We will try to get a room at the Paramount Hotel across the street.”

“Don’t worry about that,” the VC said, “I will call and make sure they have a nice room for you. The manager is an old friend of mine and owes me.”

Upon arrival at the hotel after a quick trip through immigration and customs thanks to "our" VIP status and no bags, I found a lovely suite waiting for me and a room for the driver. As we arrived at lunch time, we ate at the hotel. I told the driver that I had a change of underwear in my briefcase for this eventuality, but I needed to get two important medications, Metformin and Allopurinol. "Do you think we can find these medicines and purchase them without a prescription?" I asked anxiously.

"Don't worry," assured Kofi. "If you write down the names and the dosages, I can buy them. I have plenty of money, so I will tell you how much they cost when I return from the druggist. I suggest you get some sleep; I know you flew all night?"

"Thank you! I do need a little sleep. Let us meet at dinner. We can eat by the pool about 7PM." Off Kofi went as I headed to my suite for desperately needed sleep, setting my alarm for 6 p.m.

A Brief Overview of Ghana from 1470

Ghana (formerly called the Gold Coast) has a long history of external interactions both with Europe and through the trade routes from North and East Africa. The first recorded contact with Europe was in 1470 when the Portuguese landed and later created a coastal fort at Elmina to protect and enhance their trade in gold and other commodities in 1482. The British and Dutch also set up trading forts on the Ghana coast. A trading agreement between the British, Dutch, and the Ashanti[1] assured peace between them for a while.[2] By 1902 the British fully controlled

[1] The Ashanti kingdom grew to become an Empire through merger with other Akan people and conquest. Their army was noted for its effectiveness and was well armed They spoke a common language Twi, with a constitution, well-formed judicial system and a centralized political system based in Kumasi from 1670. They managed to keep the British out until finally defeated by them in the 1820s. Their wealth came from gold, controlled by the Asantehene (chief), slave trade of defeated soldiers, and other trade.

[2] In 1821 the British Government took control of all the existing forts along the coast and declared parts of the Gold Coast a Protectorate. Soon after, in 1844, the British established a treaty of cooperation with the rulers of the Fanti Kingdom which controlled the area along the coast—eager for allies to protect them against the Ashanti in the interior, who had better weapons and an efficient army. The Ashanti also controlled most of the gold mines. Several inconclusive wars transpired between the British and the Ashanti Kingdom with heavy losses on both sides. The British finally defeated the Ashanti armies in 1874.

what is now Ghana including the North. The Ashanti leader, called the Asantehene, and his entourage had been exiled to the Seychelles in 1894.

One curious feature of British treatment of the Northern Region of Ghana was based on their undying respect for traditional leaders. This fostered agreements with the Northern Chiefs while simultaneously maintaining the area in its "uncorrupted" traditional status.[3] Schools were not built in the North until 1908 and then solely the "sons of chief" were allowed access designed to produce a group of educated chiefs literate in English. Some students were sent South for secondary school. Most of the chiefs feared that harm could come to their sons once they crossed the Yeji River on the way South. Thus, the first few students sent South to school were the sons of slaves in the guise of sons of chiefs, with one or two exceptions. Some young men were sent to Cape Coast for primary education in the South and later to Achimota School in the capital Accra.

One unusual aspect of that "liberal" education policy (as the British saw it) was that the Northern boys (no girls were sent), were expected to leave their school uniforms on the Yeji Ferry on their return to the North for vacations so as not to corrupt the Northern population. British colonial political leaders feared that too much contact with Western education would "spoil" Northern students. As one Provincial Commission in the North put it, "To give these primitive children a more advanced education would be a doubtful blessing at present. It might make them discouraged with their lot" (Thomas 1974, 439).[4] Even by 1919, only four Government Schools were open in the North with 203 men and 8 women with classes only through Standard III (3rd Grade US) in contrast to 14,000 Ghanaian students in the coastal primary schools alone at that time, most with a full range of primary classes (Thomas 1974. 439). Fearing spreading demands for independence, as the British were seeing in other colonies such as Algeria and India, the British government made major efforts to keep Western ideas out of the North to avoid the disturbances

[3] That is leaving the existing systems of chiefs in place and not trying to unify their structures as in some other places.

[4] Thomas, Roger. (1974), "Education in Northern Ghana, 1906–1940: A Study in Colonial Paradox," *International Journal of African Historical Studies*, vol. 7, no. 3, pp. 127–164.

already experienced in the South. The North was neglected for so long given its lack of resources, there were no such problems at that time and the British hoped to keep the North that way.

Ghana experienced constitutional government for a number of years before independence with a Legislative Council as early as 1850.[5] The British responded rather graciously to demands for independence and moved to establish an independent Ghana on March 6, 1957, with Kwame Nkrumah as President. This was a period when both he and Ghana served as models for much of the rest of Africa and several other parts of the world because of the easy transition to independence, parliamentary rule, a relatively free press, expanding good road networks, agricultural extension, and a growing education system. The roads helped move the coffee and cocoa to the coast for sale abroad, and the economy thrived.

After an impressive start with national development, a free press, and democratic government, Nkrumah and the Convention People's Party (CPP) government became increasingly authoritarian and lost much of its popularity. President Nkrumah was overthrown in 1966 by a military coup, followed by power struggles within the military and growing corruption.

A series of counter-coups followed in April 1969, January 1972, 1978, and 1979. Another coup took place in December 1981 followed by civilian rule from then on. Amazingly, in Ghana these military interventions did not lead to ethnic or regional conflicts, which were often a consequence of military coups elsewhere. However, they did lead to a great deal of corruption by the military, and national hostility to the military in politics. Ghana is one of the few Sub-Saharan African countries that has not suffered from a civil war, and both its traditions and cultures reflect the long history of relative peace.

[5] The Legislative Council was to advise the Governor. By 1946, the British abandoned their "official majority' and allowed Ghanaian's representatives to be in the majority. This served as the basis for an eventual Parliament in 1950.

University for Development Studies (UDS)

The University for Development Studies (UDS) was established in 1992 after demands by Northern Members of Parliament (MPs) for a university in the North. UDS was allocated limited funds and given a few abandoned government buildings (mostly failed technical schools) near the regional capital of Tamale, in which to start classes. The University benefited from the generosity of local chiefs, businesses, and local governments which provided funds and helped rebuild and expand the existing facilities. UDS became a major local project—which continued for years including while I was there during 2002 and 2003.[6] My effort was to help the Institutions of Higher Education prepare a strategic plan that would gain the approval of the Carnegie Corporation. The Carnegie Corporation promised up to one million dollars in aid for UDS if it could produce a high-quality strategic plan. The stakes were high!

That first evening in Accra, Kofi thankfully returned with my medication—a dozen of each of the two pills. In the meantime, the hotel washed my dirty laundry, so I had clean clothes to wear the next day. They did the same the next day. Happily, my suitcase arrived the second day at about 9:00 a.m.—early enough for us to reach Tamale and perhaps Navrongo that night. Kofi was an excellent driver although his average speed was 90 miles per hour on a paved straight road, which nonetheless worried me—though I did not say anything. I was confident of his driving while scared someone else would hit us. He made the nine-hour trip in seven hours.

March 2003

[6] The North was the poorest part of Ghana, with a malnutrition level in 2011 in which "almost two in every five children are stunted and more than 80% of children suffer from anemia" (UNICEF Ghana, n.d., 4). The largest percentage of the poor were women, who also showed lower levels of literacy than men. The area's agriculture was underdeveloped, and in recent years had seen a decline in output (Kaburise 2003). The literacy rate for Ghana several years after the UDS was established was estimated to be 70%, while the average in the North was only 37%, ranging from as low as 12% in Gushiegu-Karaga to 40% in Tamale, the largest city in the North. Ninety-five percent of the population worked in the informal sector, mostly in Agriculture. See also, Hayward, Fred, (2017), "Lessons Learned from Strategic Planning for Improved Teaching and Learning in Developing Economies," *Planning for Higher Education Journal*, V45N4.

The University for Development Studies

The strategic planning committee (comprised of faculty, students, and administrators) started writing the Strategic Plan for the UDS in early March 2003. UDS was a unique institution with a curriculum that included a "third semester" in which teams of students in Medicine, Social Sciences, Agriculture, and other subjects worked with a local community to produce and implement a development project—including raising the funds for it—over their undergraduate careers. That had resulted in broad support for UDS and higher education throughout the North.

Now the University was in the process of laying out a five-year plan for the institution as a whole, and I was there to facilitate the planning and writing. The Committee and I worked together, quite effectively, for about ten days with great comradery, occasional laughter, and good progress. The English language proficiency of educated people in the North resulted in the best English spoken in Ghana, in part because there are more than a dozen ethnic groups with language mostly mutually unintelligible, so they speak to each other in English all the time. In the South, while they are taught in English in school, they speak to each other outside school and even on the playgrounds in their native languages of Ga, Twi, or Fanti. Common knowledge of English made our work go faster at UDS than if I was working in the South where a translator might have been needed.

March 19, 2003

Thoughts a Few Hours Before an Expected US Invasion of Iraq

On this day, in March 2003, the mood was somber at the UDS in Northern Ghana. My normally polite and laughing colleagues on the Strategic Planning Committee were quiet. They were mostly UDS faculty members from Northern Ghana wearing suits and ties; a few from the South and administrators were similarly dressed, and the two students on the Committee also wore Western attire without ties.

The choice of clothing contrasted drastically with that of the rural northerners who wear robes if they are Muslim, but not much of anything outside the cities if not as adults except occasionally around the waste. Indeed, in the far North at that time in Fra Fra country, men and women went naked other than using a piece of leather animal skin on their backs if working in the fields to protect them from the sun. The young girls and boys wear no clothes. The girls do wear a "chastity belt" of beads around their waist which does not cover any private parts to my surprise. I thought that was because the original belts were made when they were younger, and as they grew taller took the belts with them. I never asked. Later, one old woman volunteered how they worked. The strings on the beads are purposefully weak. If the young girls were up to no good or just broke the beads by accident, the girls would go running to their mothers with the broken beads in hand. Their mothers could instantly tell if sexual activity had taken place or not—and usually sex did not she explained.

Over lunch in a corner of the meeting room, away from me, a discussion started about the potential war the US was about to start in Iraq, which they heard a great deal about on the daily morning BBC broadcasts, based on the claim that the Iraqi army was building weapons of mass destruction (WMDs). I could hear the hostile tone of the conversation. The committee members knew my feelings—soundly against this action in Iraq because I did not believe the argument about an arsenal of WMDs.[7] I asked to join the group explaining that I was eager to hear their wisdom. They agreed. The faculty members in particular were the angriest. At least half of them had studied in the US, and all regarded my country as a beacon of hope, democracy, and reason. They had seen the horrors of war from refugees from neighboring countries, especially French atrocities during the colonial era in the Cote d'Ivoire, of a sort they did not experience from the English, or the horrors they saw from a number of refugees coming to Ghana from neighboring civil wars in Liberia. The Ghanaians were horrified by the prospect of this big power attacking a smaller country.

My colleagues felt personally insulted by the expected US invasion and talked about how wrong and unfair this was. As the Deputy Vice Chancellor (DVC) said, "Most of us regard President Bush as a bully, a cowboy, under educated, and an ideologue. We question his motives and feel helplessness in the face of naked power."

One of the students added, "We feel our inability to do anything, even though we know this invasion is wrong—the opposite of what the US,

[7] Colin Powel, former Secretary of State, years later confirmed to me that nuclear weapons did not exist—that he was duped by the President and others. Personal communication in 2005.

which we have often loved, and almost always respected, stands for." They then asked me, "How do you feel being an American at this moment?".

I told them, "I am embarrassed, sad, discouraged, and disheartened. We live in a democracy, and I am on the losing side in this case. Like those who prevailed in the Bush administration, I will have to suffer the consequences of a decision made by those elected democratically." Yet, while my Ghanaian colleagues understood my opposition and knew I agreed with them, they were almost apologetic for their anger—nonetheless, they were angry. And so was I—and at that moment embarrassed as an American.

These were the intellectuals of Ghana. The feelings of the general public were less articulate although not dissimilar. They all listened to the radio, usually the BBC, and were amazingly aware of daily events in the world. As the Vice Chancellor said, "What is clear is that the US has succeeded in raising Saddam Hussain to the equal of Bush. What we would love to see is putting Bush and Saddam in a ring to fight for a solution—winner takes all. We feel strongly that Bush should go on trial for war crimes. The Vice Chancellor reminded us, "The US has abrogated the treaties on the environment and refused to sign the land mine treaty. The US is against the World Court and has abrogated many of the civil liberties Americans hold dear. We have also seen the victims of land mines planted by opposition forces when refugees from Liberia fled here to Ghana for medical treatment and know of the damage wars create."

Outside the University, anger was also widespread. Not an anger that I feared, rather one I understood and shared. One problem was, as one Muslim chief told me, "The Ghanaians feel helpless and cheated by the United States, dismissed, and demeaned—like the rest of the third world. Half of the North is Muslim and half either Christian or animist,[8] yet we all coexist peacefully. We regard the US as a whole (though not all Americans), as anti-Muslim. Is this true?".

"No, it is not; rather there are groups who foster that attitude, and a few in Government, including Bush, who share that view, sadly," I replied.

I also received this kind of question in many of the villages I visited during my work where the UDS had development projects underway, as well as in Tamale at my hotel. More than hostility, people were puzzled since to them the action seemed so unlike the US, they thought they

[8] Referring to traditional religions. These varied, often by region or ethnic group.

knew, and the questions were more seeking an explanation than showing hostility toward me. Nonetheless, this was a challenging question to answer honestly, since I believed most people in the US, in fact, looked down on Muslims.

I responded, "We have certainly discriminated against Muslims prior to this crisis. Bush and his conservative supporters seem convinced of their religious righteousness against Saddam, Ben Laden, or others of the Islamic right. Yet we have a significant Muslim population in the US—and have for generations—many born in the US. The current situation is a sad fact of our failure to integrate effectively, understand Islam, and a tendency for bias against people of different backgrounds."

I expected the anger would intensify as pictures of bombings, dead civilians, and destroyed cities filled the airwaves. No amount of righteousness would overcome that. We will have an obligation to pick up the pieces in a context, I feared, in which the US will celebrate our "victory" and belittle those with doubts and those who criticized our actions. I hoped that neither the US cemeteries nor those in Iraq would be filled with people in the numbers that marked the war in Vietnam. None of us needs another black wall with names of the dead. This was a turning point for the image of the US as far as Ghanaians were concerned. I expected we all would live to regret this action. I hoped I was wrong.

August 3, 2003

Thoughts Over Dinner at the Accra, Ghana Paramount Hotel the Night Before Travel

I was thinking of my family while sitting by the swimming pool in Accra, Ghana (the capital of Ghana), with a marvelous band playing music to dance to. I enjoyed an excellent meal of "cuda" and "palava" sauce and a couple of glasses of wine.

A light rain created a wonderful pattern of circles in the pool while I was protected by a huge umbrella. This hotel poolside was a perfect place to people watch—those who thought they were important, those who were invisible, those who were in love, and those who wished they were elsewhere. Lots of other men were also dining alone looking uncomfortable, constantly circling the veranda with their eyes. I felt for them since I did not feel alone. I could understand how the "women of the evening" earned their living in a way I had not before, since most of the men looked so sad and lonely. The bar by the pool was filled with such women nursing drinks, provocatively dressed, and smiling demurely at anyone who caught their eyes. For me, the uniquely quiet and lovely evening was in many ways romantic—a night to enjoy alone, not wishing I were elsewhere, and not feeling sorry for myself. Of course, I was not alone, with thoughts and memories of my family for company. Plus, I was feeling quite content with the work we accomplished on this day.

Earlier in the evening, I enjoyed the gym, and the community of work friends during the day. Now, I relished this beautiful moment relaxing alone by the pool thinking of my loved ones in the States. I would sleep well that night.

January 2003

Traditional Healing in Ghana in 2003–2004

I was working and teaching in Southern Ghana on a Fulbright Grant for a year at the University of Ghana, Legon—one of the best and most active universities in Africa. My project was to study political attitudes and experiences that affect those attitudes, focusing on five different villages in various parts of Ghana. One of these was Apiride, a community about sixty miles North of Accra in the hills beyond the fertile plain. Apiride was well known for having produced the priest who helped unite the great Ashanti Empire. Ghana was a major producer of coffee and cocoa resulting in a successful economy as long as the price was stable, which in recent years it was. Apiride benefited from productive coffee farms which were reflected in that most of the homes were brick with tile or tin roofs, and the roads were paved. Two schools operated in Apiride, as well as three Christian churches for the 1400 residents.[9] Electricity was on its way with the polls already in place and the houses attached and electrified before I finished my project. Apiride was also recognized for the spectacular ritual healing powers related to broken bones, barrenness, various sicknesses, and mental illness.

While carrying out a survey there, I learned that several of these cures were not working effectively lately because a dispute between clans in Apiride prevented appointment of the fifth priest (one for each of the five clans) which meant that major healing rituals could not be performed. I

[9] At this time most of the South was predominately Christian of various denomination while the North was about half Muslim, one-third Christian and the rest traditional religions. Many Christians and Muslims also follow traditional religious customs.

was staying in a compound near the house of Chief Saffu,[10] chief of all Apiride and a great deal of the surrounding area including several villages. His second wife was cooking for me for a small fee. He was required by tradition to take a second wife after he became chief and have a child by her, a process he did not want to undertake, so his wife chose the second wife (a friend of hers from schooldays). Ghana was divided into hundreds of chieftaincies, each chosen locally from a chiefly family. Sometimes there was only one such family, although usually there were several chiefly families who rotated the chiefdomcy or from whom the chief was chosen by the elders, on occasion leading to conflicts about whose turn it was or the appropriateness of the choice. That had been the case in Apiride where five such clans existed, though they finally had agreed on the current Chief about six months earlier.

One morning Chief Saffu, who became a friend, called me to his house—a lovely brick house part way up a hill, looking down on the city. As Apiride did not yet have electricity, at this point, the chief had built an elaborate battery system and charger which he attached to his car for an hour each evening, so they enjoyed dim light at night. His driveway was paved up to the house. The Chief shared the house with his older brother and wife and three children—one by the second wife as required by traditional law. He could not have been fully recognized until she produced a child fathered by the chief.

When I arrived, Chief Saffu reported, "Do you remember what I told you about the tensions that prevented us from appointing a priest from one of the clans? I have finally succeeded in getting agreement to appoint the fifth priest. Now we will be able to carry out the ceremonies needed to solve several problems in Apiride. We are going to deal with one of these later in the morning. I am inviting you, with agreement from the others, to witness this ceremony in the sacred bush since you have shown us great respect for, and interest in, our traditions. I will pick you up in an hour, and we will walk through the forest to the sacred bush for the ceremony."

During my interviews, I learned that the inability to carry out these and other ceremonies became such a divisive factor in the town that several people sought to remove the chief (called destooling). The current chief, Saffu, was recently installed and was working tirelessly to bring peace and

[10] Not his real name.

cooperation to Apiride. Getting the fifth priest appointed was one of his early successes and had taken months of negotiation with two of the five clans.

Thus, later that morning the first of these ceremonies was to take place in the sacred bush—an enclosed area in the forest of mahogany, cotton trees, and wawa tree (a hardwood). The shrine had been used as long as anyone could remember for ceremonies and could only be entered with all of the priests' present. The ceremony would create the conditions to deal with many challenges. Now that there were five priests, they would have the powers to cure which had been absent for several years. Chief Saffu noted that of particular concern to the towns' people was the number of women who were barren and needed a special ceremony and drink to become pregnant. He said I was the first foreigner ever accorded the privilege of watching this process. I felt immensely fortunate to be invited, since I knew several other scholars who worked in Apiride and were not given the opportunity to participate in a similar experience.

Chief Saffu and I met and walked on a barely visible trail into the sacred bush—a dense forest beneath which grew banana palms and coffee in profusion in the shade of the larger trees—indeed coffee and cocoa trees needed shad to thrive. We continued for about twenty minutes to a walled area and were admitted through a wooden gate by one of the priests. Inside were several tall trees which were left because of their beauty and as a sign of fecundity, more than 100 years ago when this place was built. Otherwise, the area was cleared of plants. The enclosure also contained a small hut and large stones set out as chairs.

The chief priest came and shook my hand saying, "You are welcome my friend. We know you can be trusted and that you are interested in what makes things work well here in Apiride. For that reason, we wanted you to witness this important process. As you can see, there is a dagger stuck in the tree with a bag of herbs hanging from it. No one can pull the dagger out except the chief priest and only when five priests are present."

Several other priests tried to pull out the dagger and failed. The Chief Priest turned to me saying, "Now it is your turn to try to pull the dagger out of the tree."

I responded nervously, "I believe what I have seen of the other attempts and would rather not try (fearing what would happen if I succeeded and the others were only pretending)." The Chief Priest then proceeded to pull the knife out easily and removed the bag of herbs and other ingredients.

The Chief Priest opened the bag and removed vials of powders, leaves, and various liquids. He mixed all this in a calabash adding other ingredients—items provided by each of the other priests—to make a thick liquid. He then sampled the potion. Satisfied with the product, he declared, "This powerful medicine is ready for use. We will all taste the medicine and then give some to the barren women. Many of them had been waiting for years for this fearing the consequences of failing to conceive. As you may know," he said to me, "lack of a child is grounds for divorce. None of us wants that."

The Chief Priest passed the calabash around, and we all drank. The liquid was sweet, rather pleasant, and went down easily. The Chief Priest and Chief Saffu explained that the four women concerned were waiting outside the compound. He added, "They will be pregnant in a matter of months."

The calabash was then taken outside, and each of the four women drank of the potion. Indeed, in three or four months all of them were visibly pregnant. While all the priests were educated men, belief in the power of this ceremony was strong and universal in the area.

When telling of this experience to my students as an example of beliefs in traditional societies, I usually stopped the narrative here. Invariably a student would ask, "What is the trick?" I always responded, "I don't know; however, my guess is, as we experience in the US for couples who try and fail to get pregnant. When they adopt a child, within months the wife is pregnant. When people are tense about conceiving, the sperm and ova count decreases. When you adopt a child, you relax. In Apiride, the people involved believed in the ceremony and the medicine, so they relax, and pregnancy occurs. This has worked for years in Apiride which provides further credence of the worth of the process and strengthens the belief in its effectiveness."

February 2003

Broken Bone

The other medical specialist in Apiride, in addition to the priests and mental health society, was noted for his ability to repair broken bones (the "bone doctor"). Most of these specialists were from families that carried out these healing processes for decades and learned their crafts from parents, grandparents, and generations of relatives. This was true of the priests and priestesses who dealt with mental health as well as the bone specialist. I made an appointment to interview the bone specialist one morning. We sat in his compound amid many plants, dried leaves, and bottles of liquids made from herbs he collected. I asked him, "How did you come to be a bone specialist?".

He replied, "My father and grandfather, great grandfather, and beyond that were all specialists in broken bones. My father started training me at age five by giving me a cigar box full of foot bones. My task was to put them together properly, learning to fully figure how the bones were connected to each other. The idea was that I should learn to put the bones together, even with my eyes shut. This took me a week to learn. My father then gave me a box of hand bones with which to do the same. A few months later, my father began to take me with him into the forest to gather the herbs used to kill pain and to encourage healing, making sure I learned that well. By the time I was ten, I was helping him with patients. When my father became too old to continue with this work, I took over. We never had a case we couldn't heal."

I was delighted by his openness to all my questions and struck by the herbs he has collected—each of which had a purpose he described to me. I was also struck by how similar his training seemed to me to be to that of my daughter who was a doctor trained in surgery. I asked the specialist, "What was the hardest case you ever encountered?".

He replied, "A few weeks ago, a man was brought in whose foot was run over by a truck loaded with stones at a construction site. The foot was completely smashed, and he was in great agony. His fellow workers carried him to the local hospital in the next town. The doctors there took one look at his smashed foot and said they could not do anything for him other than amputate. Nonetheless, the doctors at the hospital suggested that the workers bring him to me, which they did. I gave him strong pain killers and then went to work on his foot. The foot was a terrible mess, and I took a while to lay out a plan. Then I worked for two hours to get the bones back in place given that many of them were smashed and in many pieces. He was in great agony even with my medication, which helped somewhat. I would not have been able to reconstruct the bone fragments without the pain-killing herbs because he needed to remain perfectly still for me to work on the foot. Even then, several of the workers had to hold his leg and arms to keep him from moving. I drew on the early training my father gave me with the cigar box of foot bones. The whole procedure needed to be done by my sense of touch.

Once the bones were back in place, though most were broken or smashed, I wrapped the foot tightly and put a caste on. He is much better now. The patient is coming in later today for me to remove the cast if you want to talk to him."

"Yes, I would really like to do that if you can let me know when he is here," I replied to the "bone doctor" (as he was called locally).

"That is fine. You will see that the foot is healing nicely when I take the cast off, though it will still look terrible and black and blue. He is now able to walk almost normally and soon will be able to walk with ease when the muscles build up again."

Joining them later in the day, I watched the removal of the cast. I am squeamish; the procedure was hard for me to watch since he was still in pain though given more pain drugs. While the damage was still clearly visible and terrible to look at because of all the bruises, the foot was obviously healing, and the man was able to walk with a slight limp which went away in a few months. The success of the "bone doctor" was amazing!

The picture of Apiride portrayed here is not unlike that of the four other communities I examined in detail during this same research as well as in most of the rest of Ghana. The conflicts in Apiride would have derailed stability in many of the other places I have worked. In Ghana, however, it is the high level of political awareness, broad public participation throughout the country, and agreement on rules for dealing with conflict that has led to such long-term success throughout the country.

Mental Health

The third area of health that Apiride was famous for was mental health. The mental health process involved a group which put people in trances to cure them of mental health problems. It was headed by a chief priest, but the real power and action was undertaken by five women priests who dealt with the individuals involved, interviewed them before the ordeal, performed most of the ceremonies leading to the trances (along with the chief priest), and "deprogramed" the individuals after the ceremonies. I attended one of these ordeals, including witnessing the interview of the candidate (done on blue books with great care) prior to the ordeal, and then attended the ordeal, the creation of the trances for both the person involved and the priests and many of those watching and dancing for at least two hours barefooted on somewhat rocky soil. I had to fight hard to avoid going into a trance myself (and a friend had warned me). The drums beat in a certain rhythm which is known to cause trances and I almost succumbed but fought hard against it fearful of what I would do if I let myself go. The woman involved believed she was a devil and had killed several people in Apiride, who had actually died. The chief believed that her involvement was all in her head. I interviewed her the day after the ordeal, and she said she was fine—and that continued for the year I worked in Apiride. They dealt with a wide range of mental health problems and seem to have a very high success rate.

Conclusions

It was the sense of community, peace, and problem-solving traditions that one sees throughout Ghana, that has helped make Ghana the peaceful success it has become with no civil wars yet successful change over the years at all levels. That was also reflected in the openness I found throughout the country—a sense of confidence and pride—also clear at the University for Development Studies—that sets Ghana apart from most of the other fifteen countries in which I have worked in Africa and Asia. While the traditions vary from region to region, they are mutually respected and honored. No one, including the Ministry of Education, tried to stop UDS as it initiated a new totally different curriculum. In addition, as we saw, local communities in the North recognized its importance and contributed both to the institution and eagerly participated in the new student and community-oriented curriculum. All that has created a prosperous, open, and friendly society. That is also part of why, in recent years, it has been such an import place of return for many people of African descent from the United States, Canada, and Europe. Its openness to other customs and traditions is a hallmark of Ghana's successes. Ghana has been a successful democracy for many years with strong public support, active citizen groups at the community level, as well as active political parties in Parliament and in the communities. It is an excellent example of how leadership at many levels with active community participation can bring about change while assuring national stability and harmony.

Especially important to its successes are the activities of local people all over Ghana, their identity as Ghanaians, their faith in government at the local, regional, and national levels. All this has led to the continued development of democratic norms throughout the country, national stability, and economic development. There is a saying in Ghana that every woman in Ghana is a "market woman," and to some extent that is true. That is part of what keeps democracy going in that everyone is involved, the economy thrives even in bad times, there is national stability, conflict resolution without violence and without the civil wars that have affected so many other countries in Africa and beyond.

CHAPTER 3

Madagascar: An Island Nation Seeking to End Its Isolation and Update the Educational System

January 2005

The Minister of Education of Madagascar

I met Haja Nirina Razafinjatovo, Minister of Education, at the World Bank in Washington, DC, in early 2005 at a meeting to discuss the needs of higher education in Madagascar. The Minister earned a Ph.D. in Mathematics from the University of Connecticut at Storrs, and later taught there and in Florida for several years. He was totally fluent in English. We seemed to have similar ideas about the development of higher education and similar values in general. As it turned out, during a break in our official conversations, we discovered that we were both huge jazz fans.

Toward the end of the meeting at the World Bank, Minister Haja Razafinjatovo asked, "Fred, can you come to Madagascar to help us to launch a strategic planning process? You are the person we need. I have already reviewed your vita. We will require a year of your time. I also want to begin accreditation and need help on quality improvement in general."

At this point we engaged in an extended animated discussion with three other World Bank officials. They were concerned about my French (which I had not spoken in 25 years), but I replied in basic French to satisfy them. The Minister interjected, "In any case, I want us to work

F. M. Hayward, *Fostering Institutional Development and Vital Change in Africa and Asia*,
https://doi.org/10.1007/978-3-031-04364-2_3

mostly in English, since I am planning for the Ministry staff to increase their English fluency as part of bringing them into the modern world."

I replied, "I would be delighted to join you, if the World Bank approves." (Somehow the Bank had already approved my appointment as senior advisor to the Madagascar Ministry of Education). I was soon headed to the capital city, Antananarivo, for my first month of work. In 2005, my wife was teaching at the University of Wisconsin, and all four of our children were off at various universities studying law, medicine, film, and acting.

Haja was a tall handsome man with an easy smile, obviously an activist and a doer. He was outgoing, yet quiet, with a certain level of charisma which made him an appealing leader. He was an expert at budgets and computers—an exception in Madagascar. Haja loved music, played the guitar with some skill, and loved to sing. He did not have a particularly good voice, but he enjoyed himself. His staff always encouraged him to sing—which he appreciated. Haja and I periodically went out to dinner or to hear some of the wonderful jazz that could be found in small bistros in the capital city.

Haja was popular with the employees, though some of the chancellors feared him because of his agenda for change, and their concern that they could not please him. This was partly, I learned later, because the chancellors lacked much power as they were elected—not appointed; thus, they generally did not want to rock the boat before the next election by implementing major changes which faculty members might not appreciate.

Haja told me, "When I came back from studying and working in the United States, I was appointed Minister of Telecommunications and did my best to upgrade the system and bring us good quality Internet. I met Marc Ravalomanana (later President) when he was thinking of running for President. He was an especially successful businessman who wanted to raise the quality of education, improve the economy, and end the rampant corruption in government. He asked me to run his election campaign, which I did. He won and largely succeeded in ending corruption. Then he tapped me as Minister of National Education and Scientific Research to carry out reforms in all levels of education. I see him regularly, and he has helped increase the Ministry of Education (MoE) budget though it still remains far too low. Now, at least, we will be able to hire new faculty after a dozen years of a freeze."

Haja and I worked to set up university accreditation in Madagascar and to encourage faculty members to raise the quality of higher education. Haja said, "The biggest problem is that the faculty members think the content they are teaching is fine. They don't know how far behind they are in comparison to the rest of Africa—and way behind expected levels of quality internationally. With your help, we are progressing."

The Minister endeavored to encourage teachers to understand the degree to which they needed to improve—in his nice gentle way—one that stimulated a positive response. In the long run, faculty members strengthened their commitment. He sent a few professors to visit other African countries, who returned in shock at the extent Madagascar was educationally inferior at that point. As they reported back to their colleagues, they created powerful incentives for change and appreciation of the intentions to develop a strategic plan for higher education.

In my two years of working with Haja, he never raised his voice in discussions, but made his case in a soft, clear, confident voice which usually carried the day. When his efforts did not gain traction, we would discuss different approaches to the issue, as we did with accreditation for the private universities, whose leadership argued, "Let the buyer decide." When we met with the private institutions again, the Minister emphasized that obtaining accreditation would be a smart advertisement of their quality. By the end of this second meeting, all but one of the private institutions realized the merits of the process. Several private colleges immediately volunteered for review for accreditation knowing that the imprimatur would differentiate their institution positively from others, if approved. The process was well underway when I left Madagascar with a number of both public and private higher education institutions having been accredited. Haja achieved many similar successes in his thoughtful, bright, creative manner.

Madagascar: A Brief Introduction to Pre-Colonial and Colonial History

The island of Madagascar presents an interesting history and culture coupled with myths and mystery about its earliest inhabitants. Today's population is basically of Polynesian origin with language similarities to that in Borneo. The first inhabitants are thought to have arrived between 200 and 500 BC by outriggers with stops along the way in India and East Africa, according to genetic evidence. DNA suggests that the early arrivals survived the 4500-mile journey in good health and were later joined by migrants from East Africa. The myth, when I first visited Madagascar, was that there were no human inhabitants on this large island 200–500 years ago. Given the evidence of extensive human activity in Africa for thousands of years, as well as major migrations by land on both coasts of Africa, it seemed unlikely to me that there were no earlier inhabitants.[1]

Particularly interesting is that the language of the whole island today is closely related to the Austronesian language family native to Polynesia. The population of Madagascar is genetically distinct from other areas of Africa, with about half the genetic material from Polynesia and the other half a mix of genetic material from East and South Africa with some evidence of settlers from North Africa and India. East Africa migrants appear to have arrived in Madagascar beginning about the ninth century according to genetic material, although for reasons that are unclear, the Asian culture dominates the population.[2] Substantial Arab trade took place from the 1600s most likely establishing small trading settlements along the coast.

By the Middle Ages, a number of separate groups of settlers formed communities ruled by chiefs in various parts of the large island. They were often at war with each other and sold their captives to Arab slave traders, many of whom ended up in South Africa. Several local leaders worked to unite the various groups into distinct entities—such as the Merina, Baklava, and Betsimisaraka—to strengthen their rule and improve their ability to trade with Europe. The kings and queens of Madagascar signed treaties of trade and friendship with England, France, and the US starting in the early 1800s.

[1] Recent archeological finds suggest that human inhabitants settled in Madagascar as long ago as 5000 years, although others disagree and suggest it was not until between 1100 and 1900 years ago when substantial numbers of inhabitants came primarily from what is now Polynesia. Bone fragments with cut marks are the evidence for the earlier estimates. However, some scholars think those are teeth marks from animals and not humans, especially since no pottery or other evidence of human activity on Madagascar island has been found to date prior to 2000–1500 years ago.

[2] Wellcome Trust Sanger Institute. (2005), "Human Inhabitants of Madagascar are Genetically Unique" *ScienceDaily*, downloaded September 16, 2019, from https://www.sciencedaily.com/releases/2005/06/05068102652/htm.

A number of powerful queens ruled between 1800 and 1897.[3] Royalty included Queens Ranavalona I, II, and III. The queens were sometimes dominated by their male prime ministers, at least one of whom was French, but most were powerful in their own right. The importance of queens exerted a powerful effect in fostering gender equity in much of Madagascar which demonstrated a relatively high level of equality among the sexes—at least at the elite levels of society in the early years of colonialism. The introduction of European ideas and Christianity helped bring matriarchal rule to an end, though it still lives on in Malagasy traditions.

Pirate activity was rampant for part of this period along the coast of Madagascar preying on European and other shipping, thus attracting British and French war ships. Later, the British and French competed for influence in Madagascar, and each country created settlements along the coast. The French eventually colonized Madagascar after an agreement with the British through the Berlin treaty of 1887.[4] The colonization process did not go easily and involved several years of war between the French army and well-armed local armies which had extensive fighting experience gained from battling among each other.

The French army finally conquered Madagascar in 1895 followed by large numbers of French settlers and businesses. The Malagasy[5] Monarchy was allowed to continue until 1897 following the French conquest. Both Christian missionaries and Muslim leaders operated in various parts of Madagascar for many years, often in conflict with each other. By the time I arrived in 2005, about 50% of the population was Christian, half Catholic, and half Protestant; with about 5% Muslim and the other 45% indigenous religions.[6]

[3] There is a long history of matriarchy in Africa, including Madagascar. Its origin is said to come from the nomadic life in which men went out to hunt or fight while women stayed and ran the community and thus became the political and social leaders. See Kama, Lisapo ya. (2020), "The African Matriarchal Tradition," downloaded on March 31, 2020, from http://en.lisapoyakama.org/the-african-matriachial-tradtition/.

[4] Ibid., Wellcome Trust Sanger Institute. (2005).

[5] The country was called Madagascar, the people Malagasy.

[6] Unlike other parts of Africa, Malagasy traditional religion is largely similar throughout the island, with a belief in a single supreme figure. They trace their unique religious customs to Asia, the Arabs, and Africa. The traditional religion is centered on one individual, called Zanahary, sometimes including a figure regarded as his son, called Andriamanitra. The supreme figure is seen as neither male nor female. Zanahary is believed

Eventually Madagascar gained independence from France on June 26, 1960, although the French continued to maintain a military base there, and a large number of French citizens continued to control much of the economy. The goal of local leaders was a democratic government with elections held regularly. Over the next decades, a number of regimes proclaimed republic status. The First Republic ruled from 1960 to 1972, the Second Republic from 1972 to 1991, and the Third Republic from 1991 to 2002. In 2001, Marc Ravalomanana's, *I Love Madagascar Party* won a decisive majority of votes and ruled until 2009 when he was overthrown in a coup. The country then fell into rapid decline with the military and police becoming extremely corrupt, seizing cars and other property from the population. These circumstances cut investment and tourism, as well as donor support. The country has not recovered from the disaster of the coup, although recently a new democratic election has brought hope of reform.

Gender Equity

The Madagascar constitution (Article 6) forbids discrimination by gender, giving men and women equal rights. The constitution decrees equal participation in the economy, government, and society for all, regardless of gender, religion, or origin. Women are also able to own land, which is unusual in many developing countries. The level of equality for women is among the highest of any of the more than one dozen Asian and African countries I have worked in over the years. The level of gender equity is highest among the well-educated and more affluent, in the urban areas, and lowest in isolated rural regions. About 80% of women work in the labor force, including farming. They own property, run businesses, vote, and serve in high offices. Twenty-three percent of the members of Parliament are women.[7] On the other hand, women's salaries are about 20% lower than men's for the same job at the lower levels such as clerks, though nearly equal in areas such as education and government, which pay comparable wages for men and women.

to have extraordinary power to reward those who do well as well and punish those who do not.

[7] Razafindrakoto, Gaby. (2016), "Gender Equity in Madagascar," *World Policy*, downloaded September 17, 2019, from https://worldpolicy.org/2016/03/15/gender-equity-in-Madagascar/.

Malagasy college-educated women are striking for their confidence, willingness to question actions they disagree with, and hold their own in work. Many grew up in a one-room house and learned how to be self-sufficient in their own space. They are also quite comfortable with their bodies and neither hide their figures nor flaunt them.

Problems of inequality for women are more severe in the rural areas where men still dominate. Many women are relegated to household activities such as growing small gardens, crafts, and raising lots of children. An indication of employment problems for uneducated women was the high level of prostitution, visible and apparently legal in most cities. Especially attractive young women waiting for clients lined whole streets.

Among the successful professional women, Felana Razafindrova[8] comes to mind. She worked for a foundation which provided funding for various development projects. She was 30 years old and managed a multi-million-dollar budget, a staff of six including men, and traveled around the country, alone, inspecting projects while hiring and firing staff. Felana took great care to find the right people; and regularly checked on those employed, firing any staff who did not work hard or were involved in corruption such as bribery, theft, or pay-offs. She had one child as a single mom, and no one questioned its propriety, though traditionally girls were married at puberty—which still occurred in some rural areas. She loved to dance and adored jazz, like most Malagasy.

Another example of a successful woman was my assistant at the MoHE, hired by the Minister as a civil servant to translate and work with me, Nirana Ramanantoanina.[9] She was a charming woman of 25, a recent graduate in English, Nirana was an amazing simultaneous translator. Nirana could translate continuously for eight hours a day (unlike the usual two hours) if necessary and was an excellent assistant. She was especially attractive with a delicate face, light brown skin, beautifully thin, and always well-dressed. She would sit slightly behind me at meetings and translate quietly in my ear. Nirana was especially popular with the Ministry staff and attracted lots of suiters. Her simultaneous translations meant that we did not have to interrupt meetings for translation, as was

[8] Not her real name. All Madagascar names are long, some up to 36 characters. The tradition for this is that for a long time Malagasy used only one name and the name was to reflect their origin, parents, and history. The more important the person—the longer the name.

[9] Not her real name to protect her under current circumstances.

the case with most translators, thus keeping pace with the process. Her data analysis was also excellent, and she brought a keen understanding of the educational process including the problems with the admissions examination system. Nirana was invaluable to the success of our work and to the success of the Minister and the goals of the President to improve the quality of education.

About ten months after Nirana started working for me, she asked if she could go to the Ministry of Finance, "I need to get my back pay for two weeks of translation work I did for the Ministry a month before you arrived. I have checked with the Higher Education Finance Office weekly, and they cannot find my records of that work. They therefore called the Finance Ministry and were told that no one there knew anything about the funds. Originally, I was advised that the Finance Department would send the money here to me through the Finance Office at the Ministry of Education."

"I agree, Nirana. You should absolutely head over there and take care of this problem. Let me know if you need anything from me," I added.

"No," she replied somewhat embarrassed, "I shouldn't have any problems." And off she went, perfume wafting in her path.

Nirana was back in our Ministry office in an hour and a half, visibly shaken and her face flushed. She plopped down hard in her desk chair.

"What happened? You look shattered. Please tell me what took place. How can I help?"

Nirana recounted an appalling tale:

"This trip to the Ministry of Finance was terrible. I ought to have taken you up on your offer to come with me. When I finally arrived at the front of the line, the clerk, after spending about ten minutes searching through a bunch of papers, came back and said, 'You have already collected your money.'"

"Impossible," I said angrily. "I've never been in this Finance office before."

"You signed for it. I have your signature," the clerk replied brusquely.

"Let me see the signature," Nirana said. "The clerk then brought the receipt book in which someone signed as 'Nirana Ramanantoanina,' though not anything similar to my signature which I showed him."

"The signature looks OK to me," the clerk replied and called over another colleague who repeated the same dismissal of my claim of falsification.

"I request to speak to the manager," I said appalled.

The second man replied, "I am the manager! If you do not stop complaining, I will have you arrested for fraud."

"At that point, I was about to break down in tears, though I refused to let them see my feelings and left quickly. I found a waiting taxi, locked the door, and came directly back here."

I tried to respond in a quiet, soothing, voice which belied my intense anger, "I am furious! If I thought you would encounter problems, I would have gone with you. We must find out who these employees are and see that they are fired and get you your money!".

At this point Nirana started to cry quietly. My instinct was to hug her, although I realized that would be inappropriate in Madagascar, so I handed her my freshly poured coke.

I said, "Clearly, there is no point in returning to that office."

"No!" Nirana shouted. "I will never go to that place again."

"We should talk to Haja, now that he is Minister of Finance, and seek his help with your funds. He has been fighting against corruption. He is adamant about creating a positive environment for women, in the Ministry of Finance as he did in the Ministry of Education." I immediately called Haja's assistant and asked if we could come to his office to discuss an urgent matter. She said, "Yes."

Over his years as Minister, Haja worked diligently with considerable success to clean up corruption[10] in the Ministry of Higher Education, especially in the admissions process, but also in examinations and grading. He was starting to do the same in the Ministry of Finance after he was appointed Minister.

We went straight to Haja's office. Nirana related her story to Haja. He was furious, calling in his Deputy, asking him to find out who took her money. He assured Nirana that she would receive her pay. The Deputy Minister was equally horrified and vowed to get answers and act against the perpetrators at once.

Haja called us back to his office a few days later to report total failure. He was upset and embarrassed by his lack of success. As he said, "The code of silence prevailed! We will keep our eyes on these people, but for

[10] For many years before the Presidency of Hery Ravalomanana, Madagascar was ranked as the most corrupt government in the world by the World Bank and others. After President Hery Ravalomanana's cleanup, Madagascar did not even make the list of the top 25 most corrupt nations.

now we are helpless without further evidence." He pulled out his wallet to pay Nirana himself.

I stopped him. "Never mind, I will take care of this," which I did. I knew Haja's salary amounted to about one-fifth of my World Bank salary, as were all Ministers' salaries as part of the President's austerity campaign. Haja put the staff on notice that he would be watching for similar transgressions.

At this point, to break the tensions, Haja offered us some of the delicious chocolate chip-cookies he always kept on his desk—a love he picked up in the US which his wife happily continued. Haja commented, discouraged, "You would think getting rid of corruption such as this would be relatively easy—especially in a case in which the fraud is so clear, and we have a forged signature. Yet, we do not know which of the possible five people in that office is responsible. They all deny any knowledge of the loss of funds. We don't have any handwriting experts to check whose penmanship fits the fake signature."

"We understand the problems, Haja," I said. "This has no doubt been going on for years, and these people are especially astute. They found a check that had not been picked up for a few weeks intended for an employee whom they did not know, so they cashed the check. No doubt, they have a deal with a worker at a bank to help with that too."

"Yes, Fred, I am sure you are correct, but we will get to the bottom of this fraud eventually, I can assure you! In the meantime, I hope our efforts have curtailed the corruption a bit," said Haja. "It is always the ordinary people who get hurt. No affront intended to you Nirana. Nonetheless, even with your connections to me, we don't have much power as we have seen—even mine is more limited than I thought!" In the long run, the Assistant to the Minister identified the culprits a few weeks later, thanks to a tip.

The President of Madagascar was close to Haja who had organized his successful presidential campaign. We met with President Ravalomanana frequently. After great achievements in higher education, Haja had been appointed as Minister of Finance to clean up that agency and try to obtain additional donor funding for education and other development projects.

There were many women involved in the process of improving higher education, in the Ministry and at the universities. In several cases they were among the leaders of change and gave great support to the Minister because of his defense and demand for gender equality. Though there was a long history of women in power, from time to time it was necessary to remind institutions of both the constitutional and education commitment to equality. Many of these were very powerful women who were successes on their own and had learned how to move the system forward. Without them, the early successes at improving the quality of education, gender equity, and greater admissions to education, would not have happened.

French Interference in Madagascar

A few months after I started working as advisor to Minister Haja, I received a hand delivered letter from the French Embassy "summoning" me to their headquarters. I was surprised by this summons and asked the Minister, "Do you think I should I go? I work for you; no one 'summons' me except you."

To my surprise, the Minister replied, "You should go. What the French have to say should prove interesting. I know they are not happy with me; they think I encouraged President Ravalomanana to make English the third national language along with French and Malagasy. They may think you were involved in that decision—which is not the case. Don't let them force you to speak to them in French as that will put you at a disadvantage. Speak only in English. Use Nirana Ramanantoanina if you don't understand. I know you speak fluent French; but they invited you there, and you should speak English as a matter of principle. The staff at the Embassy all know English. I will be especially interested to hear about this meeting."

Nirana and I arrived at the French Embassy at the appointed time but were kept waiting for more than fifteen minutes. I said to my assistant, "They are purposefully late to show us how important they are. If they do not come in five minutes, we will leave."

"The French here are always arrogant." Nirana said. "I worked for them for six months teaching English. I finally quit. I was tired of being frisked and yelled at each time I entered the Embassy. I know their condescending attitude was partly because I am Malagasy and a woman."

Indeed, we were about to leave when we were invited in to meet with the Cultural Attaché, M. Rebate, and his assistant, M. Depecker. The men greeted us in French. I replied in English introducing my translator,

"You can talk in French, since I understand French and have a translator. I will speak in English, since I know you understand my language."

"We would rather converse in French," they replied. "Nonetheless, English is acceptable."

M. Rebate started the discussion, "We read your recent report to the World Bank. As a result, we are concerned that you did not highlight French aid given to Madagascar."

I replied, "That was intended as an internal document about the World Bank's own programs in Madagascar. I didn't mention any of the other donors either such as the British, German, UNESCO—even the United States."

They then came to the real reason for the "summons" to the Embassy, "We are also concerned that you are an American. The advisor to the Minister of Higher Education has always been French since the establishment of higher education in Madagascar in 1955. The current advisor should be a Frenchman. We hope you understand our position and will work with the Minister to change the situation. After all, Madagascar is within the French sphere of influence."

"That is not my decision to make," I replied. "I was selected by Minister Haja Nirina Razafinjatovo and the World Bank staff. You would have to discuss this matter with them. I would not be involved in such a decision." I was able to maintain my temper after years of experience. I knew if I lost my temper, I would lose the battle. Nonetheless, these were outrageous requests, most of which only Minister Haja could answer. However, I was damned if I was going to show anything but surprise at their requests.

The Cultural Attaché continued, "The French plan to work here over the next five years on improving the quality of higher education. We are providing 100 scholarships for Malagasy students. We want to encourage young people to earn the qualifications necessary to be hired by the universities given the current preponderance of aging faculty due to a recent budget freeze. In addition, we have been coordinating aid with all NGOs and governments operating in education in Madagascar and want to continue this involvement. The French Ambassador is eager for us to continue working with the Minister. Prof. Hayward, we hope to have access to all the funds France paid to the World Bank in our annual commitment, to use for our projects here," he added.

"That is not the way I have seen the World Bank work in my ten-years of experience with the organization. They use those funds for their chosen projects," I said. "You need to talk to the World Bank representatives about access to funding."

Monsieur Rebate expressed concern, "The World Bank's Plans might differ from the plans of the French. We are worried about that," he emphasized. "What are your goals for the strategic plan for higher education?".

"We are working in general on quality improvement," I crisply replied, hiding my annoyance. "However, the plan is still in the discussion and drafting stage. If you want more information about the goals, I suggest you talk with the Minister."

At this point the Attaché said, "We want to assure you that we are interested in cooperating with the World Bank. We will send you a report about French cooperation with Madagascar."[11]

We returned to the Ministry to report to Haja. He laughed, "The French are terrified of US influence here. Clearly, they see you as a threat. They don't understand that President Ravalomanana and I consider the US model better than the French—as emphasized by the fact that most leading world institutions of higher education are in the States. Curiously, today I was disinvited by the French from a conference in Burkina Faso where I was to present a paper. I am certain that cancelling my presentation is intended to show their anger at the addition of English as a national language and their assumption that I am behind move. They still think we are their colony."

"That is amazing!" I am also on the program for that conference in Burkina Faso, but recently received a letter of confirmation. I suspect they did not dare cut me out because I was invited as a World Bank employee and the Bank is covering half the costs of this event. "You know," I mused, "the Attaché and his assistant were amazingly open about their need to control higher education. You warned me that the French acted arrogantly, but I didn't think they would lay their claims out so blatantly."

Haja, Nirana, and I laughed, and the Minister congratulated me for my successful meeting at the French Embassy.

[11] The document never materialized.

Music Is the Soul of Madagascar

The people of Madagascar love music[12] like no other place I have visited. Many people play an instrument—the guitar being the most popular, which is used primarily to play traditional music. Some of the traditional tunes also have group dances associated with them, called *Afindrafindrao*. When the opportunity arises, the seats are cleared, and everyone dances—from age six to eighty—usually with partners in a kind of line dance—partners can be two females or males if not enough people of the opposite sex are available.

Hundreds of traditional songs are sung by almost everyone in beautiful polyharmonic fashion. These harmonies are unique to Madagascar and appear heavily influenced by two hundred years of church music, the harmonies of South Africa, and contact with Arab traders—all melded into beautiful melodic tunes, especially practiced by the Merina people, and then spread throughout the country. Most of this music was in major keys. Jazz was also especially popular, with several jazz artists becoming so successful that they moved to Paris, performing there regularly as well as in Madagascar. A number of jazz bars and restaurants are found in the capital, Antananarivo, and in most of the major towns around the country. The music venues were always packed, as was the bar and restaurant near the University of Antananarivo which featured regular groups that played during lunch and dinner. One group consisted of three women singers with especially lovely voices. The crowd would join in singing along to many of the songs. The trio was usually accompanied by piano, guitar, and muted drums.

One evening I attended a particularly wonderful demonstration of musical Madagascar. Several of us from the Ministry went to hear a concert at a nearby military base featuring two well-known singers, Bodo and Rija Ramanantoanina, with a terrific backup band consisting of piano, two guitars (one base), and drums, playing a mix of traditional songs and jazz. The huge hall was a gym packed with at least 400 people sitting at

[12] Malagasy music is very melodic and is quite distinguishable from African music on the continent. The music is usually harmonic drawing inspiration from the Austronesian origins of the people, South African traditional harmonies, and church music going back to the mid-nineteenth century. In later years, Malagasy music has also been highly influenced by American jazz.

tables with waiters serving drinks and snacks of kabobs, fried shrimp, and sambos.[13]

The audience was mostly young, about 25 years on average, with a smattering of small children and a good number of couples in their 30s and 40s. A little more than half of the audience were women, mostly in their twenties, most in couples. The audience knew almost all the songs. The performers, men and women, were wonderful, especially singing duets. They played to the audience, encouraging them to sing, occasionally inviting people up on the stage to join them—including the man next to me who turned out to have a wonderful baritone voice. Many songs were performed without audience participation. Nonetheless, a few in the audience joined in as they knew and loved the songs. In some cases, the audience sang the chorus and the singers the verses. Almost all the songs were in Malagasy, though a few were in English or French. Some were translated for me by colleagues, but I had picked up enough Malagasy to understand the general subject and direction of the songs, and in any case enjoy the magnificent music.

A general feeling of community was generated. The harmonies were especially beautiful. The male singer, with a deep melodic voice, captivated and moved the audience. The women showed an amazing range and could belt out a song in the finest diva tradition. We left toward 1:00 am, but I am sure the bulk of the crowd lingered until the wee hours of the morning as was typical.

The next evening, I took two of my colleagues to another jazz spot which I was told featured excellent music. The entertainment was performed by a group of four women—one a well-known singer with several CDs. All were in their 50s or older, and some of their voices were past their prime—but the feelings and charm were, if anything, even increased by age. The "theme" of the day was "Music of the 60s"—mostly in Malagasy—a language that lends itself to music. The music moved along quickly with solos and duets, as well as performances by the whole group.

[13] Madagascar sambos are among the most popular snacks on the island. They are small, crunchy, deep-fried dough covered outsides with a corn-based crust and filled with vegetables, ground meat, chicken, or other fillings—all in a savory sauce.

As I sat in my apartment writing one Sunday morning shortly after our evening listening to the quartet, I could hear the choir in the Christian church next door singing another enchanting traditional song. The lead singer exhibited an extraordinary alto voice, while the other singers provided a harmonic background with an amazing fullness to the music. The service was mostly music—a combination of hymns I knew and traditional Malagasy songs. The singing was all in Malagasy with the congregation joining in on most of the songs—a joyful morning of lovely music and awesome harmony. Music is an integral part of Madagascar's culture—an aspect of their life which touched me deeply and sometimes moved me to tears for its beauty, sincerity, and comradery.

Malagasy pianist, Jeannot Jabeson, who then lived in Paris, came to Antananarivo to put on a series of concerts. I took my assistant and the Minister to hear him. Jabeson's jazz piano was especially outstanding, and we could understand why he was a recording and concert success in Europe and Africa. We heard excellent renditions of all the standards, including marvelous versions of tough pieces such as Brubeck's "Take Five" with its difficult key changes, as well as traditional American spirituals.

One has to search this enormous island to find the best Jazz since there are many potential venues. We also heard wonderful vocalists, especially women, singing in the Ella Fitzgerald tradition or offering her songs with their own take. Combining the beauty of the countryside with the warmth and gentleness of the people, makes Madagascar a music lover's paradise full of joy and human warmth.

12 November 2005

A Relaxing Afternoon of Food, Conversation, and Jazz in Diego

A few days later, we attended the inauguration of an impressive computer project at the University d'Antsiranana in Diego, which included 80 computers, five servers, printers, and other equipment donated by a Swiss company. Then, the Ministry staff took us to a well-known dining spot outside Diego for the most wonderful chicken on the entire island. Four of us from the Ministry plus two newspaper reporters, drove out the main road to the edge of town where houses and fields were intermixed. We arrived at a traditional African hut with a cone-shaped thatched roof, a rondaval, called *Tonton Benavony* which means "uncle's improved moved village."

A retired army officer ran the restaurant out of the small straw structure, with food eaten outside at tables with tablecloth, napkins, silverware, and glasses all set for us. The chicken was marinated in several herbs, garlic, oils, and other secret ingredients which were especially delicious. The chef is said to start the process of preparing the dish only after guests arrive—killing the chicken and preparing the dinner in under 20 minutes. The dinner was out of this world. The proprietor must have been 70 and did most of the serving himself, with his grandson helping when he arrived back from school about halfway through our dinner. We drank beer, coke, and other soft drinks. The Minister and I downed Three Horses Beer—an excellent Malagasy brew. Plates full of chicken kept on coming with that wonderful marinade that was a bit spicy and sweet at the same time. Everyone was in a joyful mood. Minister Haja joked with his staff, and conversation continued in a mix of French, English, and Malagasy.

From the *Tonton Benavony* we returned to the hotel for a 20 minutes "rest," and then were driven to a jazz club. The venue was a beautiful new hotel, Villa Samantha, on the outskirts of Diego with a scenic view of the Indian Ocean and the hills. We were ushered into a dining room, open to the water on one side, where a band was playing a Ray Charles piece. To our surprise, another meal had been prepared with great care for the "honored guests"—our third meal in six hours. This time, we were served excellent Madagascar large grilled shrimp, more chicken, and fresh fruit salad for dessert.

The high point of the evening, however, was the incredible band. Better than anything in Tana claimed my table partner, a journalist working for national television reporting on the Minister's trip. She spoke excellent English, which helped our interaction. While I could understand almost everything in French at this point and was once fluent, my ability to speak with ease was limited—though in a pinch I muddled through.

I knew Minister Haja was a great jazz fan, familiar with all the famous pieces. The two of us evoked lots of smiles as we swayed to the music and identified almost all the composers. We heard Brubeck's *In Your Own Sweet Way* and a whole host of other familiar jazz tunes, plus a range of Latin pieces including Jobin and others I did not know, that I thoroughly enjoyed. Everyone was dancing, and I was even taken out onto the floor by one of the reporters. She was a great dancer and absolutely lovely. The group tried to force me to sing with chants of "Fred," "Fred," "Fred," but fortunately for them I resisted.

I engaged in a number of fascinating discussions with several of the people in our group at the Villa Samantha Hotel. The delightfully relaxing day and evening led me to suggest to the Minister that he move the Ministry from Tana to Diego. My comment pleased our host, the President of the university, an impressive leader and a delightful woman. She was a linguist trained in Madagascar with post graduate work at UCLA. The President was well regarded for her creative ideas, a tremendous amount of energy, and an extraordinary intellect.

2003–2007

Flora and Fauna of Madagascar

One morning Nirana and I drove to one of the national parks dedicated to the protection of rare plants and animals, especially as a sanctuary for Madagascar's indigenous and world-famous Lemurs. The experience was unforgettable! We saw more than ten types of Lemurs, thanks to my colleague's sharp vision and the forest ranger's attraction to my outgoing assistant. He thus gave us the cook's tour of the area, including going deep into the thick bush, cutting our way with his machete, to see the pigmy Lemurs. More than fifty types of those wonderful primates exist with their human-like big eyes and hands. Since they have no natural enemies, they were fearless, until the Chinese, new to Madagascar, began eating them. The Lemurs came right up to us, eating bananas out of our hands, even while carrying babies on their backs. Some varieties were as small as softballs; others stood more than three feet tall and moved by jumping on their hind legs.

Madagascar boasts an astonishing range of plants and incredible orchids, with a wide range of bright colors, which grow in abundance on trees in the protected forests We saw more than one hundred orchids as we walked through the park. We also spotted beautiful species of birds—plumed with bright feathers of yellow, red, and green.[14]

[14] Sadly, since the coup in Madagascar in 2009, the wildlife, plants, and trees have been plundered by poachers, lumber companies, and the Chinese—with bribes paid to the army, police, and forest rangers. The new government was not serious about protecting its natural resources, unlike its predecessors. Several species of plants and animals, including

Madagascar produces more vanilla than any other place in the world. Vanilla is a species of orchid and must be hand pollinated to produce commercial quality products. The actual pollination is usually carried out by young girls with fingers small enough to complete the delicate process successfully. The vanilla which had been harvested that year did not sell as ships were afraid to dock in Madagascar due to violence after the coup. We visited several farms, which were beautiful with rows and rows of vanilla plants. Sadly, local planters were suffering because they were not able to sell their vanilla.

Crocodile Lake[15]

During our exploration of the northern part of the island, we were taken by Diego University leaders to visit a lake, which was of great significance to those living in that area. The lake was about 250 kilometers south of Diego where we were staying, and a dozen miles from where we participated in the inauguration of the new classroom blocks. Our entourage traveled in a caravan on dusty roads with one or two sweep cars ahead to provide security for this quiet, unostentatious Minister, Haja Nirina Razafinjatovo.

At a rest stop, the Minister steered me back into his air-conditioned car. We headed off quickly on our own with our driver and one security person to gain a head start and avoid the inevitable dust cloud churned up by the sweep cars. Passing the sweep cars was a smart move since the route was especially bumpy, rough, and dusty. The trip was breathtaking, if harrowing, with expansive views of miles of rice fields in a valley below. In the distance, stood red colored bluffs and cliffs lit up by the sun. To our delight, the threatened rain dissipated, and the sun emerged turning the whole scene into magical hues of blues, green, and gold. The sunshine accentuated the iridescent light green of the young rice seedlings recently transplanted and thriving in the early seasonal rains.

As we approached our destination, we spotted the lovely, large deep blue lake off to our left with no houses or people nearby. Indeed, the one person we saw was riding a bicycle weaving back and forth in the

several types of Baobab trees and some types of Lemurs, have already become extinct over the last few years.

[15] A traditional story as told during a visit to the Lake.

dust, with a huge load of rice and firewood, causing our driver to swerve precariously to avoid him as we passed.

We stopped in a clearing near the edge of the lake. A retaining wall prevented cars from approaching the water's edge. We noticed a fenced-off area with rough cut posts about one foot apart. Inside the enclosure was a shrine with four flat stones and a pole in the middle onto which bulls' heads were stacked with the pole running from the bottom of the skull through the top. More than a dozen bull's skulls were staked on the pole with horns all aligned to about 10 feet high. Next to the pole, at least a dozen more skulls, in various states of decay, lay on the stones. Evidently, some of these skulls had been laying there for years. As we walked closer, we could smell the awful odor of blood from a freshly slaughtered bull. Indeed, closer inspection revealed that this sacrifice was recent, with the animal's intestines visible a distance away in the bush along with other evidence of a feast and celebration.

By this time, the others caught up with us. One of the teachers told us the story of the site and the reason for the shrine and sacrifices:

She began by telling us that a long time ago there was a village near this spot which lacked a ready source of water—the lake did not exist at that time. People in the village walked several miles for water and carried it back to the village. One day, an old lady came to the village and stopped at the first house she saw to ask people for a drink of water. The people refused, saying water was too hard for them to get to share. The old lady continued to ask at each house along the way and was refused by everyone. Finally, she reached the last house in the village and again asked for a drink of water. The woman inside gave her the last water in her house. The old lady thanked the woman and told her to leave the village immediately and move to higher ground, because something bad was going to happen to the rest of the villagers. The woman did as she was told.

Shortly after the old woman left, the valley filled with water, destroying all the houses, and engulfing all the villagers except the woman who was warned to flee. The people themselves were turned into crocodiles. Indeed, one man with a wooden brace on his leg has been seen as a crocodile with the brace still on his leg. People now come to this exact spot to feed the crocodiles and seek help with their problems or fulfillment of their aspirations.

The local people understand that this is a sacred lake. They will not wade into the water or disturb anything in or around the lake. Nevertheless, strangers ignore the warnings and bad things happen to them. For example, some foreigners brought a speed boat to the lake and were water skiing when their motor stopped. The skier managed to climb back into the boat, but the driver could not start the engine even after trying for hours. The two men were not able to paddle to shore. Someone at the shrine eventually heard their cries for help. He told them he could free them to get out of the lake if they agreed never to return. After they promised to respect the lake, he poured some libations on the stones at the base of the shrine. Immediately their motor started. The visitors left and were never heard from again."

We were told that people, though Christian, still believe in the power of the lake and came to make sacrifices to thank the gods for recent successes—often to fulfill a promise to make a sacrifice if they succeed—or because they thought an offering would bring them good luck and well-being in the future or help with a problem. They slaughtered a bull, added that contribution to the pile, and shared the meat with the crocodiles. The crocodiles are treated with respect as if they were people. Fishermen have great success at the lake catching good-sized fish. However, they believe in only taking the fish they can eat themselves or give to friendly neighbors, since selling these fish would bring bad luck.

While we were at Crocodile Lake, the twenty-five people with us spoke in quiet tones and maintained a respectful distance of 15–20 feet from the edge of the water. As we looked out on the lake, several of the local teachers started clapping in a slow rhythm and told us the crocodiles would come to the water's edge at a clearing near the shrine. Sure enough, one enormous croc came right up to the edge while another swam over from the other side of the lake. They were offered some meat. We watched for a while and then went on our way, happy to let the former denizens of the village—now crocodiles—have their lake and water to themselves.

Inaugurating New School Rooms South of Diego (Antsiranana)

About 60 kilometers south of Diego, in the far North of Madagascar, is a small town which was celebrating the opening of two blocks of new classrooms at a secondary school. The work was paid for by the Ministry of Education for one block of classrooms, and the World Bank for the other. The buildings were impressive structures with several well laid-out and stocked classrooms in each. The furniture was donated by local businesspeople—in one case impressive, enameled wood with metal reinforced legs made in China.

The ceremony was a whole community affair. We arrived through a flower-covered archway to long lines of school children in blue or green uniforms on both sides of the street, clapping and cheering. I wondered how long they had been standing waiting for us. I was traveling with the Minister. When we exited our 4 × 4, we were greeted by schoolgirls who put garlands around our necks and exchanged the usual French kisses on both cheeks. We then proceeded to a covered platform with mic and speakers. Several hundred students, teachers, and dignitaries were sitting in a half circle around the stage. The mayor, the regional police inspector in his white uniform, the directors of the schools for the area, the CISCO chiefs (school districts), local businessmen, civic leaders, the president, and vice president of the university from Diego all attended this joyous event, plus an assortment of national and regional Ministry people and press who traveled with us.

The speeches were in Malagasy, but Minister Haja sitting next to me on the dais kindly translated the critical parts—including letting me know when I was being introduced and when to stand up to take the applause and shrieks from the students. The speeches were a bit long, but the crowd appeared attentive, and the master of ceremonies was incredibly engaging. He managed to keep the folks alive, often laughing and encouraging the students to cheer, respond to his queries, chant, and enjoy the celebration. I was asked to speak too and kept my remarks very short suggesting the students work hard so they could attend the university. The Minister translated my oratory into Malagasy for the audience.

The students were lined up in their various uniforms and were quite disciplined—amazing for me to observe and they clearly followed the speeches. This was an especially fine secondary school with student entry by examination. Most of the students would do well enough on the admissions examination to attend a university if they chose to. Indeed, when the Minster asked who wanted to go to the university, most raised their hands. He also asked how many could use a computer—and about 25% responded in the affirmative. He then asked if they knew about technical college, a concern of the Minister's regarding a lack of enrollment in those institutions—a handful responded positively.

Following the event, about 200 guests were invited to a lovely lunch laid out in a rice storage shed nearby where long tables and chairs were set out. The meal was excellent—rice, tender beef, and potatoes served in a tangy sauce, accompanied by sliced cucumbers in oil, bread, and soft drinks. As is traditional, we ate with our hands, and everyone looked at me at the outset to see how I would do. Fortunately, I had eaten in this manner in India many years previously. The group clapped when I used my fingers correctly without getting sauce all over my clothes or up to my elbows. One is not supposed to get food beyond the first knuckles—mine were sauce-covered up to the second knuckles. The luncheon was served by young high school women with great care, overseen by two older women who were the efficient and capable organizers.

The people of the North of Madagascar are especially handsome and beautiful with slim figures, high cheekbones, light brown skin, and lovely faces. I encountered more attractive people in one day than I had seen in Antananarivo in three weeks. The atmosphere was particularly open and friendly with a relaxed air and dress that reminded me of Southern California where I grew up. People were clearly delighted that we were present. Minister Haja gave a wonderful, animated talk that drew responses and applause from the students. The day reflected examples of people working hard to make a difference, to improve the quality of education, and to build better lives for the young people of Madagascar.

Attempts to Overthrow the President: Thoughts on the Conflict in Madagascar

At a rally held by the Mayor of Antananarivo, Andry Rajoelina (whom people called the DJ, since he was an extremely popular music host), at the *13 Mai Square*, the Mayor announced that he was leading a movement to overthrow the President, Marc Ravalomanana (called "Dada" by many). The DJ, the Mayor, announced that he was prepared to step in as the country's new president. The DJ argued that the incumbent President violated the constitution by closing the DJ's TV and radio stations because they broadcast a message from the previous president, Didier Ratsiraka (who has been convicted of a variety of crimes and fled to exile in France to avoid prison). The DJ accused the President of acting in a dictatorial manner and failing to understand or respond to the needs of the average citizen. The Mayor asserted that people were suffering from increased food prices and inflation of other commodities, limited employment opportunities, and the growing economic crisis. His speeches drew huge crowds, spurred on by his ardent supporters who then marched through the streets of the capital city.

Some in the crowd, (I am not clear if they were encouraged directly by the DJ or by his supporters), burned down the national TV station, destroyed the studio of the national broadcasting company, incinerated a number of businesses owned by President Ravalomanana (an especially successful businessman before he became mayor of Tana and then President), and broke into and looted a number of other businesses. Demonstrators could be seen carrying bags of rice, boxes of goods, and even furniture from the looted shops. Others were selling stolen goods openly on the street. While the size of the crowd was hard to estimate, certainly thousands participated as was clear from the footage I saw. An observer could sense a party atmosphere from the facial expressions of the rioters. Many were apparently paid by the DJ, along with the French who were appalled at President Ravalomanana's open-door policy to other European and American countries.

For the most part, police and security forces[16] did not intervene. President Ravalomanana later argued that human life was more important than property, and he was not prepared to see people killed in an effort to prevent the destruction of property. Nonetheless, more than 25 people died in the carnage, most apparently, when the roofs collapsed in the burning buildings as they were looting. The press reported that the army, police, and fire department stayed out of this chaos to remain neutral. Historical precedent for their neutrality existed, but I am not certain that was the case. After the first day of violence, however, police were present in Antananarivo, as were the gendarmes who were out in great numbers, guarding government buildings, businesses, and the oil refinery.

The DJ held several meetings, appeared on private TV and radio, and continued to demand that the President resign. He also sent a letter to the Constitutional Court in Antananarivo demanding that they force the President to leave office.

Originally, the President and members of the government told people to ignore the Mayor and go about their business as usual. Nonetheless, the city was paralyzed for a week with most businesses closed. On

[16] The police while national were for local use while the security forces were a militarized national guard (gendarmes) and the military (regular army).

Monday 2 February, a few businesses opened though major stores and shops remained shuttered. A curfew was imposed from 9:00 pm so restaurants were closed, taxis off the streets, and the capital city grew quiet. The violence spread to several other cities including Tamatave (a bastion of opposition support) more than 250 kilometers away.

To an outsider such as myself, the actions of the Mayor seemed bizarre. While he was recently elected Mayor with 63% of the popular vote over a member of the President's party, he was a political novice and merely 34 years old—thus too young to legally run as a candidate for president according to the Malagasy constitution. His demands were ad hoc and did not follow any constitutional procedure for removal of a president. His position that he would serve as the head of a "Transitional Government" sounded particularly bizarre. Following constitutional procedures, the Prime Minister and the legislature could remove the President, or the opposition parties could mobilize to demand his removal through a vote of no-confidence. Those options were not taken.

This destruction of property and the conflict itself reminded people of the crisis that occurred in 2001–2002. At that time, the previous President, Didier Ratsiraka, refused to step down after Marc Ravalomanana was elected. His refusal to relinquish the office of President after losing the election precipitated a crisis that lasted a year, divided the army, and pitted people from the interior against coastal people (the closest thing to an "ethnic conflict" and "ethnic identity" in Madagascar). That crisis was costly, resulting in many foreign companies leaving the country reducing jobs by 50,000 (which had still not been recovered) and resulting in a per capita income decline from about $550 per person to $360. Many people feared a repeat of the previous experience only several years before.

The DJ is said to have called for foreigners to stay inside. I did not follow his advice and went to the Ministry of Higher Education for my expected consultations, where I was greeted warmly by the staff. I went out for lunch in the city center without observing angry crowds or evidence of current activity, though many gendarmes casually stood around government buildings and large department and grocery stores.

The Constitutional Court rejected the DJ's request advising that they had no such power, and he had no standing to make such a request. On February 4th, the government removed the DJ as Mayor of the city, appointing Guy Rivo Randrianarison, PDS (leader of a special delegation) to act as Mayor of Antananarivo. He was accompanied by the chief of police and sworn into the office. The press, originally sympathetic to the

DJ, were now generally calling for law and order, respect for the Constitution, and generally mocking the former mayor for his attempted coup and self-serving offer to take charge. Unfortunately, the crisis was not over.

President Ravalomanana toured the country to enhance his support, and the now unseated mayor promised to do the same. The DJ reflected many people's unhappiness with the economic situation at this time, although the economic difficulties in the country had not been caused by President Ravalomanana. On the contrary, the President had measurably improved the economy with new investments in Madagascar. The demonstrations in this regard were not unlike those in Greece, France, Pakistan, and other countries against the government; actually, the peoples' cry for help at a time of inflation in which food prices increased as much as 60%, fuel prices increased, and growing numbers were unemployed. However, most of these increases were a function of world prices rising for gasoline and other imports.

For me, it seemed that the DJ was trying to use the economic crisis, President Ravalomanana's purchase of a new Presidential airplane, the President's independently earned wealth, and caution, to push his own agenda and political aspirations.

The DJ did go to court on the issue of the broadcasts but lost. In one sense, his broadcasts were exercises of free speech that should have been tolerated even if they crossed the margins of legality—and the broadcasts did, in fact, break the law. The DJ was pushing the envelope trying to mobilize people to stage a non-military coup, while perpetrating violence. No one benefited from the devastation, violence, destruction, closure of stores, factories, and offices. Indeed, these activities led to tremendous economic costs to Madagascar since then. The crisis demonstrated the delicate situation in Madagascar and in much of the rest of the developing world cutting off a main source of income, tourism as well as exports.

More Thoughts on the Crisis in Madagascar

My one-year appointment by the Bank kept getting lengthened, at first by the Bank and then by donor funds under the control of the Ministry. The political crisis continued with no end in sight. On March 1, 2009, people rallied in support of President Ravalomanana—about 40,000 people came to the stadium downtown to support him. Although rain poured during most of the event, people seemed happy and excited in expressing their support. They were, as one newspaper called them, the "silent voices" of the capital. The large turnout was important for the President and his supporters, as the DJ had succeeded in monopolizing the spotlight and the press with rallies and events.

The determined ex-mayor continued to appoint "ministers" to his "transition government" in a game that I perceived would go nowhere. Already, eight "fake" ministers had been appointed. The actual Ministers were in their offices running things. Various attempts at "mediation" (United Nations, Organization of African States, Indian Ocean Partnership) failed partly because, in reality, there was nothing to mediate as the President gained no advantage by promising new elections, and the DJ was not interested in abandoning his efforts to unseat the President. Talks continued between representatives of the two sides but were canceled after several weeks. No one could figure out how to easily resolve these issues, especially since the DJ remained determined to become the "interim head of state."

Some Malagasy sensed that mass demonstrations were a way to change governments as a kind of "mass street democracy" in action. Looking at the crowds, as I did, as they converged on the square in front of the President's Office, they included predominantly young, probably unemployed or underemployed men. Many of the marchers seemed out for a lark and clearly were having fun experiencing the revolt as a kind of game. Others were paid to demonstrate. However, many sincerely supported the ex-mayor hoping he would end the economic crisis, while others were looking for an opportunity to loot or cause trouble. Robberies and other violent incidents occurred all over Antananarivo and within several other cities.

My colleagues at the Ministry were upset and discouraged. They worried that their work in education reform would go down the drain, the economy was in free fall, and prospects for the future were growing bleak. I feared they were right. Ninety percent of tourists canceled their visits. I decided to postpone my visit to Diego in the North scheduled for the next week since I had no assurances of a flight back.

One big investor, South Korean Daewoo, pulled out of a farming scheme to grow corn. The vanilla crop could not be sold (sales were to start the day before the killings) since the port was closed due to violence in the area. Ships could not take the crops to buyers. Prices were rising, especially for rice and oil. The rice in storage had been looted. The stored rice was to cover the shortfall between the end of the growing and harvest season and the new harvest. Madagascar would need to import rice, rather than export it—which would cost foreign exchange they did not have.

In the meantime, some funders including the International Monetary Fund (IMF), and The World Bank cut off money to the government until they had reviewed the national finances. The funders were especially concerned about the government's funding of the President's airplane.

People were frightened. Many more lost their jobs before this crisis was over. The collapse of the international economy, and its devastating effect turned many people against the government and the President, even though they were not responsible for it, and had in many ways softened the blow to Madagascar compared to other African countries. Even if the ex-mayor had been able to take power, he would not have been able to improve the economy.

The damage to the nation's peaceful reputation was already undone by the populist drive and their resort to violence. Anyone with an ego such as that of the DJ would likely remain blind to the realities of ordinary people and the policies needed to steer development back on track. A charismatic leader may be able to organize the unemployed and disaffected, especially when some of them are paid to demonstrate under devastating economic conditions such as these; but putting the political system back together, as well as building the economy, requires a different set of skills and experience which the DJ did not have, and should have been obvious to many more people.

The French were partly behind this turmoil. According to some observers,[17] the French were providing money for the demonstrations. They were angry at President Ravalomanana for opening trade to non-French companies, in particular from Canada, England, and the US. They were furious that the President had made English the third language of the country. The French regarded Madagascar as "theirs"—part of their sphere of influence.

February 7, 2009

[17] See for example, "Dear France, Thank You for the 2009 Coup," *Madagascar Express*, March 5, 2012.

Thoughts on the Second Round of Violence in Madagascar

Today (February 7) was another day in which the ex-Mayor of Antananarivo called a rally—this one to announce his appointment of a Prime Minister in his new "transitional government." The DJ continued his efforts to replace the President of Madagascar in a bizarre, but deadly, effort to make himself President. The consequence of this rally was more violence than the last one with a number of deaths.

I had arranged a lunch meeting with a colleague from the Ministry of Education and my assistant at a restaurant near the Office of the President. As we were finishing our meal, the waiter came to our table advising, "You must leave now! We heard on the radio that the crowd attending the rally for the ex-mayor is headed this way to attack the Office of the President. We fear that you, Sir, as a foreigner, will be hurt by the mob if you are nearby when they come."

My Malagasy colleague added, "Malagasy usually are peaceful and gentle people—the public does not fear our people in this situation; however, prudence requires that you return immediately to your apartment (which was two blocks away)."

Indeed, some of the people we encountered as we left the restaurant were excited, looking more like thugs than political activists. Hundreds of demonstrators were already converging on the route to the President's Office. More were following behind, armed with clubs and other tools, shouting slogans, and beating on garbage can lids. The military declared the area a security zone—off limits to anyone. These angry demonstrators were not going to pay attention to that warning.

Opposition radio reported, as did Catholic radio, that the mob marched to the Office of the President, along with the person designated as the new "Prime Minister" by the ex-Mayor demanding admission to the building. They argued that since all this land was part of the Commune of Antananarivo, the people could rightfully take it over. The soldiers and officials guarding the Office talked to the leaders of the Mayor's delegation, but absolutely refused to allow a take-over of the Office of the President. We were told that a round of shots rang out, with more following—perhaps from the army or the armed supporters of the Mayor (some said by government soldiers who had joined the Mayor). Many people were killed. I saw the wounded on TV, though news coverage was minimal. Calls from friends, against the DJ's efforts, suggested that at least twenty people died.

I heard the official (real) Prime Minister, a former army general, on the news on Catholic radio that evening after these events. He decried the deaths but urged people to remain calm, and to avoid violence, crowds, and non-constitutional actions. He extended the curfew for one week stating that if necessary, he would call out the army to maintain order. President Ravalomanana also called for calm and peace.

The ex-mayor, on the other hand, declared that if he were in charge, he would lower the price of oil and rice. Lowering prices, as he suggested would require government subsidies. At that moment, the government did not have funds for such an operation. They could barely pay wages. This strategy made for great grandstanding, appealing to those who were poor and easily manipulated.

In response to the violence and confusion, The World Bank and IMF suspended their programs of budget support to the Madagascar government. The President had made purchases which called his financial credibility into question, actions which played into the hands of his opposition. One wonders why the President was so irresponsible. Yet, for all his faults, he had been properly elected in the first place, then reelected two years ago. During his two terms, he had improved the nation's economy and enjoyed strong support in Parliament. He was dictatorial and arbitrary in several of his actions recently and was less charismatic than the ex-Mayor though very magnetic and a strong advocate of democracy.

Most of the press were now appalled by these tumultuous events, urging constitutionality and negotiation between the President and the DJ. Reporters were cynical in their recent descriptions of the attempted "coup" by the ex-Mayor. The Protestant and Catholic churches sought to mediate. Those offers were turned down by both parties.

Several of my colleagues admitted they were embarrassed by the current situation. If the crisis were not so serious, the DJ's attempts would have been regarded as comic opera. But the economy was damaged by the DJ's efforts, not to mention the loss of life. His opposition to the President's actions may have been sincere, although my own view was that his attacks were personal and vindictive with strong French government backing. He must have known that his call to action would also bring out the thugs (or perhaps he organized them himself) and cause death and destruction. On another day, at least a dozen more buildings were burned! I could see smoke from my apartment. At least ten more deaths occurred over the next few days.

The closing of most businesses caused huge losses, both the salaries of workers laid off and the loss of business. The cut-off of donor support limited the government's ability to pay salaries, provide support for education, health services, etc. I figured the closing of most shops would continue into next week. The US Ambassador, in my mind, had not been particularly helpful, seeming at the outset to support the Mayor—then blaming everyone.

The situation suddenly seemed untenable with no easy solution. I did not worry about myself, but I felt devastated that all the hard work which had gone into national development would probably be undone, including our efforts in education. My friends in the Ministries were quite distressed. I assumed they would probably be okay; but the ordinary people, who suffered so much in this poor country now with a per capita annual income of $260, would suffer.

On March 17, 2009, the army turned against President Ravalomanana with a unit paid by the French leading the effort against him. The President turned over the Presidency to the head of the army and went into exile in South Africa. The army then handed over the government to the ex-mayor. Both the economy and law and order immediately went further downhill, as predicted. The former Minister of Education, then Minister of Finance, Haja, had to go into hiding since all former ministers were booked for arrest, though none were charged with a crime. After a year in hiding, Haja escaped to South Africa by ship meeting the vessel from a small motorboat beyond the two-mile limit, by prearrangement. He was later able to move to the US where he remains today, teaching at a university near Washington, DC.

The country remained in deep economic trouble in 2020, with the loss of 100,000 jobs in 2019 and a failure to resume development and education reforms started by President Marc Ravalomanana. In spite of elections overseen by the military, the results failed to produce an effective government. Former President Ravalomanana remained in South Africa in exile until 2014 when he returned to Madagascar, was briefly jailed, but released soon thereafter. The former mayor was elected as head of state in 2018, and the country continues to suffer both economically with a lack of law and order even as the DJ tries to return the country to the rule of law with a focus on national development. Those efforts have been hurt further by the Corona-19 virus pandemic.

The tragedy of Madagascar continues with the economy still in difficulty and tourism only increasing slightly due to fear of instability. Crime and corruption continue to be a problem although recent efforts finally seem to be curtailing some illegal actions. Exports, especially vanilla, have not made up for the loss of investors, few of whom have returned. The significant progress made in upgrading primary and tertiary education[18] has come to a halt with many of the improvements reversed.

In a sense, the post-coup situation is similar to that of so many other countries in which the military and/or police have intervened. There is a breakdown of law and order, and police and military use their new powers to amass wealth. This is also a demonstration of how a charismatic individual can mobilize people under difficult economic conditions, which are not primarily the fault of government, in ways that turn out to be destructive and detrimental to both the economic and political situations of the country. This is a case in which large numbers of people who were suffering economically were mobilized for the inappropriate, yet appealing, goals of a popular national figure in an effort that succeeded in removing the elected government but made the national situation worse. This is an excellent example of how public mobilization can be used in ways detrimental to the public and the nation. Ten years later, at the time of writing in 2021, Madagascar is still suffering the consequences.

The coup and its aftermath have been devastating. The country was making such major progress in improving the quality of, and access to, education, finally hiring new faculty members for the universities after a twelve-year freeze, upgrading primary and higher education significantly, though failing to improve secondary education during that period. Access was increased, the percentage of women students grew at all levels, quality was significantly improved and checked by accreditation, and private education had been brought under a degree of Ministry control to weed out the weak for-profit institutions. Corruption had also been brought under control for the first time in decades by the former President. And in the middle of an international economic crisis, the economy was doing reasonably well with the benefits broadly spread especially in agriculture and tourism. However, even those benefits were meager, and the average citizen did not understand how well Madagascar was doing compared to

[18] Secondary education was also being revised but had not made much progress by the time of the coup. It remains of low quality.

most of its African neighbors. While the President and others in government tried to make that clear, these were not issues and comparisons easily understood by the average citizen. This left the door open to a very charismatic Mayor of Antananarivo to play on their fears and to accuse the President of financial abuse by buying a new presidential airplane.

Unlike what we saw in Ghana, for example, there was not the strong political attachment to government at any level, or to political parties (as in South Africa), nor to government. That had been damaged and in many ways lost during the populist uprising created by the former mayor of Antananarivo. This was less the fault of President Ravalomanana who had cleaned up corruption, improved the economy, greatly improved government functions, but due to the long history of corruption before his term—Madagascar being ranked the most corrupt country in the world at one point prior to Ravalomanana's presidency—had left people very cynical about the national government and ready to see malfeasance and mismanagement in mistakes made. While President Ravalomanana had managed to lower its rate to around 27th that was not seen by the ordinary citizen. He had also gone a long way to build trust and demonstrate that honest government could work. It would have taken years of such work to build the kind of confidence, trust, pride, and allegiance we saw in Ghana. President Ravalomanana did not have that time. Sadly, after the coup the corruption rate rose to 155th, a low, in 2017, and was down only slightly in 2020 to 149th.[19]

Those leading the changes at every level before the coup were dedicated and hardworking. Most of them now have left Madagascar for other French-speaking African countries, the US, and France. Coincidentally, Minister Haja lives in my area, and I see him and his family periodically, now with two children. He has been asked to return to Madagascar several times by the new government but does not trust them. Haja and his family hope to visit at some point. The former President has returned, thanks to international pressure, but his actions are limited. News from my friends and former colleagues suggest that the economic and political situation has not improved much. That is a sad commentary for a country that was finally progressing even after being labeled the most corrupt state in the world.

[19] Trading Economics (nd circa 2021), "Madagascar Corruption Rank," downloaded June 12, 2012, from https://tradingeconomics.com/madagascar/corruption-rank.

I am afraid that the coup and its aftermath are a huge step backward for a country that was once doing so well—especially bad since the current government has further destroyed public support of government, trust in both national and local leaders, undone most of the great progress made in improving higher education and other areas of government. Charisma only works if those leaders show progress in the economy, stability, and personal well-being. As we have seen, that has not happened. Corruption is back with a vengeance; the military and others have been stealing people's property and amassing money in other ways. Tourist, a main source of income, have fled to other safer venues in Africa and elsewhere, the vanilla production was disrupted, major foreign investors have left, and the unemployment rate is up. Time will tell if faith in democracy and government can be restored, and the economy rebuilt.

The failures in Madagascar, as those in Sierra Leone, were partly a product of loss of public confidence in the elected government, sparked by a world-wide economic crisis, but in this case exacerbated by a self-seeking Mayor who built on that unhappiness and, with military and French assistance not only overthrew the elected government, but brought chaos and corruption back to a country that knew it well from the pre-President Ravalomanana period. Once that was done, as we saw in Sierra Leone, it is much more difficult to rebuild respect for democracy in a now corrupt system, improve the economy or continue work to improve education, health, and other areas of the system. We see again here what the loss of followers can do to even the most successful of governments.

Nonetheless, Madagascar remains one of the most beautiful places I have ever worked in having wonderful, protected forests and amazing animals like the Lemurs. The people are warm and friendly and helped me throughout my stay. And the music still plays in my mind as well on the many CDs I brought home with me. I know that is one pleasure Malagasy still enjoy. I would love to go back.

I was able to leave Madagascar a few days after the coup catching one of the last planes leaving the country at that time. While there have been recent elections, the DJ is still in charge. Aspects of life are improving somewhat, but the economy has not recovered nor has tourism returned to its previous prominence. The DJ's election to the presidency does not bode well for the future. This is an excellent example of how a charismatic leader can take advantage of a world-wide economic crisis for his own gain yet leave the country much worse off than it was before his intervention. It is also yet another example of what foreign interference and military coups can do to overturn democracies and create a broad distrust of government that is hard to reverse. The ongoing deterioration of the country has also resulted in the loss of much of the nation's top talent fearing arrest and/or seeking opportunities which will take advantage of their expertise. It also demonstrates the problems that result from low levels of public participation and a lack of broad deep commitments to democracy factors that cannot necessarily resist military interference, but could lead to major efforts to restore democracy, something still missing from Madagascar.

CHAPTER 4

A People's Liberation in South Africa

Perhaps no country has endured such a long violent struggle to establish a democratic government as South Africa.[1] That effort required not only a change of leadership, but extensive transformations in the national culture and the laws. Such radical reforms were designed to end racial inequality, gender discrimination, increase access to quality education for all races, and establish a wide range of policies to create a country in which politics were truly participatory and government was transparent, with improved quality for all. This involved major changes at the national, institutional, and for many, at the individual levels.

In higher education, the initial efforts at change took place at the institutional level, while new policies for the national level were being debated and defined. As was the case in Afghanistan, South Africa was a nation in the midst of severe conflict and violence, an internal war merely slightly less violent than that of Afghanistan, sparked by decades

[1] South Africa is at the Southern tip of Africa surrounded by Namibia, Mozambique, Botswana, Zimbabwe, Swaziland, and surrounds Lesotho. Its population in 2020 was estimated to be 59 million with a land area of 1,219,600 square kilometers. It has a very diverse land mass including tropical forests, deserts, high mountains that have snow on them in the winter, and rich farm areas, especially around Cape Town noted especially for its wine and fruits. It has rich agricultural exports, fish, gold, diamonds, and other minerals.

F. M. Hayward, *Fostering Institutional Development and Vital Change in Africa and Asia*,
https://doi.org/10.1007/978-3-031-04364-2_4

of racism, violence, and the blatant inequities of apartheid. Estimates of people killed range from 7000 to 30,000[2] during the struggle over more than forty years. The totals are debated since no one has accurate records. In addition, many thousands of people were jailed for pass law[3] and other violations of apartheid legislation, while thousands more fled to other countries when they faced arrest or intimidation.

Among the major demands of the African National Congress (ANC)[4] and other parties opposing the status quo were concerns about apartheid education policy, which required separate education systems by race and restricted access to the best higher education to white students eliminating that for black[5] students at "white" universities. During this period, blacks could no longer attend white universities.[6] The *Extension of University Education Act* (1959) decreed that separate "tribal colleges" would be established for black university students. Thereafter, black enrollment was limited primarily to separate historically black institutions. Only 9% of black Africans of college age were enrolled in higher education, 11% of colored[7] students, and 33% of Indian students, while

[2] For a discussion of the deaths and claims of deaths on the lower side, see: "I Love South Africa but I Hate My Government" (2009), downloaded on July 23, 2019, from: https://iluvsa.blogspot.com/2009/11/no-retribution-for-mbeki.html. My own view is that the number was about 14,000.

[3] Black citizens were required to carry pass books which gave their "race" and said where they could live. That was abolished when Mandela was elected.

[4] The African National Congress (ANC) was founded in 1912 as the South African Native National Congress, changing its name soon thereafter to the African National Congress. It was established to fight for the rights of the black population. It founded a military wing in 1961 when negotiations failed. Elections eventually took place in 1994 and led to a victory for Nelson Mandela and the ANC.

[5] "Black" in this context is used to refer to part of the majority population of South Africa that was discriminated against. Historically the group was divided into African, Colored, and Indian/Asian. The term "black" is also used to describe people of "black" African descent in contrast to Colored and Indian, which is sometimes confusing.

[6] A few students were admitted with permission of the Minister of Education or through other dispensations. However, by 1953 only 1064 Black African students had graduated from a university since 1652 (Moleah 1993, pp. 415–416).

[7] The term "colored" referred to people of mixed race. A recent fascinating article (Quintana-Muret, Lluis, et al. 2010–2011) has explored the origin of the Colored population. They concluded that their origins are primarily from five population groups: Khoisan, Bantu, European, Indians, and Southeast Asians. The predominant group was the Khoisan peoples (more than 60%) with an almost negligible maternal contribution of Europeans. The major group born in the early days was from "...massive maternal contributions of

60% of that age cohort of white students (Bunting 1994, p. 39) were enrolled. They were only 11% of the total population. The actual population figures of South Africans by race at that time were blacks 77%, whites 11%, coloreds 9%, and Indians 3%.[8]

The long struggle and then negotiations that brought about the dramatic transformation in South Africa result in what Nelson Mandela called a "national consensus on democracy, equality, and peace."[9] The end of the struggle led to the first truly national elections in South Africa in which all citizens, regardless of race or gender, were allowed to vote. In that context, the following account takes place.

Working with the Historically Black Institutions (HDIs)

In 1992 the American Council on Education (ACE) met with leaders of the Historically Black Institutions (HDI) and the Ford Foundation in South Africa to consider ways in which together we could be helped to upgrade their institutional quality and effectiveness given the legacies of apartheid, which was then officially still in place. The HDIs had been underfunded from the outset and in many cases their white Vice Chancellors were put in place more to keep the Black folks under control than to oversee the provision of quality education. One, the University of Fort Hare, had been established in 1916 by interested white liberal citizens, black educators, and several churches including the Presbyterians, the United Free Church of Scotland, and the YMCA, as the South African Native College. It attracted students from all over Southern Africa including Oliver Tambo, Govan Mbeki, Robert Mugabe as well as Nelson

Khoisan peoples and the almost negligible contact with respect to their paternal counterparts." The overall picture of South African Colored population today results mainly from early contact with European men and minimal contact with European women. The major contact early was with Khoisan and African males with autochthonous Khoisan females (Quintara-Muret 2010–2011).

[8] Some of this material is drawn from Hayward, Fred (2020, pp. 77–78).

[9] Parts of this discussion draws from an unpublished piece written by the author, Teboho Moja, and Nico Cloete in 1955 entitled "Change in Higher Education," which was a think piece on strategic planning through a Ford Foundation funded Pilot Project on higher education change in South Africa.

Mandela. However, in 1958, with the Extension of University Education Act, it was put under the control of the apartheid-created Ciskei Bantustan as a Xhosa-only institution like the other Black higher education institutions in the country. The Ciskei Bantustan was one of the government's attempts to establish black mini states, as were most of the other Black institutions with their separate Ministry of Education under the control of the South African Government. After the victory of the ANC in the elections it again became semi-autonomous with its first Black Vice Chancellor, Sibusiso Bengu, who would later be appointed Minister of Education in the Mandela Government.

It was twelve of these historically Black colleges, universities, and technikons that we were to work with from 1993 to 1997. We worked with the Vice Chancellors and the Ford Foundation to set up a program for their development. The Vice Chancellors wanted American University Presidents to mentor each of them. We agreed and provided them with a list of US Presidents eager to participate in the program, from which they could choose one who seemed to meet their needs. We had a mix of male and female Presidents as well as a number of minority leaders among them. Each would work at the South African institutions twice a year for two weeks two times a year—a major commitment of a month away for a sitting president. In between they would communicate regularly by e-mail and telephone. The focus would be that each of the institutions would write a strategic plan for the next five years.

What was fascinating about this program was that it brought together people who had been at the forefront of the anti-apartheid struggle mostly working in small groups, to think collectively about what future South African higher education should be like. These were leaders, faculty members, and students who had not legally been able to meet under the laws of apartheid. What we saw was the outcome of years of thinking about post-apartheid higher education while they were subject to apartheid. As someone who had studied South Africa for years, I was nonetheless surprised by the extent of apartheid impact when I started working with these institutions. For example, at the University of Fort Hare there were separate health systems by race, the best for white faculty and staff, then colored, Asian, and finally the least extensive for Black faculty members. In addition, there were separate programs for the small number of female faculty and staff for each of the four categories of people. Variations of this plan were replicated at all of the HDIs.

What was striking was that even in their isolation, their goals were amazingly similar. To be sure all wanted to see an end to apartheid, but they also wanted strong faculty governance, an executive body that was representative of the population with distinct powers that would limit political influence on the institutions, high quality including accreditation, research funding (which was largely nearly non-existent for the HDIs), better student support systems, egalitarian health care, and catch up funding to make up for the lack of investment in the past. While many of the white faculty members were strongly pro-apartheid, the hope was to integrate them, rather than remove them. Amazingly, at most institutions that worked, and in some cases, as at the University of Fort Hare, several die-hard apartheid believers soon went to the Vice Chancellor to apologize for their past behavior and beliefs and offer to do anything to make up for their past. It was these individual and collective efforts to overcome the deeply imbedded racism on campus and in the communities, that led to the successes of the ANC in transforming higher education and the country as a whole. It was an amazing transformation to watch unfold.

Request That I Come to South Africa to Help Identify and Prevent Expected Fraud During the National Elections

I received a phone call in early April 1994, while at home in Washington DC, from the Reverend Frank Chikane, the well-known head of the South African Council of Churches following the retirement of Bishop Desmond Tutu. The Government had tried to kill Reverend Frank several times because of his successful anti-apartheid preaching.

"Fred," Frank asked with the sound of urgency in his voice, "can you come to South Africa tomorrow to help us prevent the fraud we expect in the national elections? The elections are from April 26 to 29, so we do not have much time to sort this out. You will officially represent the United Nations Mission."

The Reverend recently became a member of the South African National Electoral Commission and wanted to talk about problems regarding the first democratic election in South Africa. He called after consultation with the Commission which agreed to ask me to come and assist their efforts. Many of them knew me from my work with the HDIs.

"Fred, I know you have worked on elections in many different African countries and have been working with historically black universities here in South Africa and know my country well. We are aware that our adversaries, including the Inkatha Freedom Party, the Zulu cultural group which became a political party aligned with the apartheid government, and those who support the existing Government (a few of whom are on the electoral commission staff), are trying to fix the results of this election. However, we don't know exactly how they plan to disrupt or distort the results."

I replied, "I don't think I can come tomorrow, but probably the day after. I need to get permission from the President of the American Council on Education (ACE) for whom I work, though I am sure he will say yes." As noted earlier, I had been working in South Africa part-time at historically black institutions on a higher education improvement project supported by the Ford Foundation. President Robert Atwell said "yes" and so I headed off to Pretoria, South Africa, the very next day.

On the long flight, I had time to think about Frank's earlier visit to Wisconsin in 1996. I was then Dean of International Studies at the University of Wisconsin, Madison. Frank had come to the university to see his wife who was studying African linguistics with my wife, as guest of the university. Upon Frank's arrival in Madison, I received a phone call from students who picked Frank up at the airport, explaining that the Reverend had become quite ill and was taken to a hospital. When I heard the news, I knew Frank had been poisoned given my knowledge of the anti-apartheid struggles in South Africa. I immediately went to the hospital. The doctors told me he was near death and on a ventilator. I asked if they had tested for poisons, and they said yes and found no evidence. My guess was that this was a less known poison. Frank eventually recovered thanks to their intervention and later that of the University of Wisconsin Hospital where he was transferred, yet without discovering the cause of his collapse. He was hospitalized three more times before we figured out the type of the poison at my insistence with help from a group of doctors at the Veterans Hospital near by, and then a treatment for it. The substance turned out to be a largely unknown poison at that time, thought to have been developed by Russia and known only to the US, Russia, and Great Britain as far as we knew. Once the poison was identified by some former military virologists from the Veterans Hospital—thanks to its relation to one used on artichokes in California because of its short half-life—we learned of an antidote.

We could not determine the way in which the poison was administered until years later through the "Truth and Reconciliation Commission" hearings in 1996 which were broadcast nationally in South Africa. I was listening to the hearings on my car radio when driving to one of the historically black universities (part of the Ford project on strategic planning) when the actual perpetrator apologized and explained during his testimony his scheme. I was so stunned I pulled my car to the side of the road to take in what the perpetrator was saying. He put the poison in Frank's underwear after Frank checked in for his flight to the US. Thus, each time Frank changed clothes, he was poisoned again. From Frank's point of view, I saved his life by my perseverance—a view he shared with everyone including his congregation at an emotional Sunday service in South Africa at which I was present. Several other opposition leaders were killed earlier by that same method. Now there was a known antidote.

When I responded to the request of the Election Commission that I come to South Africa to help identify the potential fraud, the Reverend Frank Chicane met me at the airport on an April day in 1994 and took me to the Carlton Hotel in Johannesburg to check-in, freshen up, and then to the Electoral Commission Office across the street to meet with my new colleagues. Frank and I talked all the way there about the problems of setting up the elections and his suspicions about those likely to be involved in attempted fraud. I told Frank and then the other Commission officials, that I needed a day to read the elections laws to identify potential weaknesses for fraud, and then I would want a tour of the preparations they had made for counting the votes and for communication with the polling places.

The Election

During my first few days in South Africa to prepare for the vote-counting process, I attended several briefings for election observers and several meetings of groups promoting election turnout and fair elections. These included several mixed-race student groups from the University of the North, WITS, and other universities in the capital, which were at the forefront of the anti-apartheid movement. I was impressed and touched by the democratic ethos that prevailed among them—a cardinal rule and basic principle of the ANC's participatory process. This shared devotion to democracy was particularly palpable at a meeting with students which included a senior ANC activist/university professor, who led a discussion on goals of the election. Each one of the dozen White and Black students present was given an opportunity to speak before anyone spoke a second time. At one point when a senior professor tried to interject a comment, he was reminded that he had spoken earlier and no one, not even a distinguished professor, could talk until those who wanted to speak were given their turn. The Professor apologized; the discussion continued. Over time I was to see this pattern occur again and again.

My review accomplished, Pirishaw Kamay, who was chosen to be my partner in the counting process, with some oversight from Prof. Francis Wilson, started to work. Pirishaw was a union leader with substantial election experience. Pirishaw was tall and good looking and understood the election process and ways to cheat from his experiences with union elections. He was born in South Africa of Indian extraction, well-educated in Durban, and broadly trusted.

On my day two, we went to look at the counting setup. One quick look at the computers and fax machines told us the system would not work though at that point we did not say anything. People could vote at any voting station in the country. Each station was given a list of all registered voters in South Africa. Nineteen million votes were expected, with results to be sent by fax to the Electoral Commission after every 2000 votes were counted, both by Fax by the local election commissioners, and also telephoned into the electoral commission by the chief observer. We would compare the two totals to be sure they were the same. The ANC made the request to count totals after every 2000 votes in hopes of avoiding fraud. This meant that in large constituencies results could be transmitted ten to twenty times. The Electoral Commission had five fax machines and ten computers—far too few to deal with 19 million votes in a reasonable amount of time.

I suggested to those in charge of the facility that we test their counting program, and they were eager to show us how it worked. I gave them made-up figures for the first 2000 votes for three imaginary political parties. Their technicians entered the data easily into the computers. Then I presented 2000 more trial numbers for a theoretical second set of results. They were not able to enter a second batch. The computer program was set up for a single entry. Obviously, the program needed to be rewritten so that votes could be added after every 2000 voters, and a record of each entry kept so that results could be rechecked later if necessary. The system clearly required more fax machines and computers to handle the load. We made calculations and suggested the additional hardware, including telephones, they needed. They had two days to get the system right.

The first day of the national elections went off smoothly with a few little hitches here and there—late opening of a polling place, misplaced equipment—eventually the problems seemed sorted out. However, late in the afternoon of the first day, the Belgian electoral police specialists sent to help us as part of Belgium foreign aid—and gladly accepted by the Commission—realized that three million blank ballots were missing and reported that they had been stolen from a "secured" warehouse. This meant the election was three million ballots short of the expected number of voters. This had to be an inside job by someone working for the Commission. They never found out who was responsible while I was there.

Pirishaw and I met with the election commission members to decide the next steps. Several commissioners wanted to call off the elections saying that the ballots were printed in the UK with special watermarks, and we could not get more ballots from the UK in time. I said "no," we can print more ballots overnight. I knew the Electoral Commission owned printers as they had used them to print sample ballots. "What we needed to do was remove the overprinted 'Sample Ballot' from the model and print three million ballots by the next day—no small task."

The Commission members continued to be worried that if we printed more ballots in South Africa (the only way possible to produce them overnight) the new forms would not have watermarks. I pointed out that only we knew about the watermarks and would know which ballots were which. If we did not say anything, those interested in committing fraud might not know we printed additional new ballots, since they would look identical to the original minus the watermarks. Thus, the original and freshly printed ballots could not easily be deciphered from the originals at the last minute.

We managed to print 4 million new ballots overnight in great secrecy using three different printing companies and the Electoral Commission printers. I organized the South African Air Force, working with a general by phone, to deliver the new ballots to the proper constituencies. Late on the second day of voting, Commission members mobilized observers for each flight, and the Air Force delivered the forms to the areas short of ballots, so that they would be available on the third day and later if needed. The all-white Air Force had insisted that the ballots be accompanied by civilian observers during the flights so they could not be accused of fraud. We provided neutral observers for each flight.

On the second day of voting, the college students we hired to work with us in the counting rooms, discovered that we were receiving what seemed to be second sets of results for the first-round ballots we had received, from several constituencies. These duplicated those we had already tallied but with different totals—these favoring the Government Nationalist Party and the Inkatha Freedom Party. We asked the helpers to pull all the original results for those constituencies and identify the fraudulent counts—a major task. At the same time, we asked our German computer expert to check the election computer; Pirishaw and I suspected altered totals, and she found an illegal software entry which added votes to pro-Apartheid parties' totals by 10% more than the actual vote at each entry. Over the next few days, with the help of a team of young men and women college students, the German computer expert, and three Belgian police who specialized in election fraud, we worked to resolve the problems we discovered. The addition of a sub-routine to change the results in our main computer was especially serious—most likely an inside job by someone working for the Electoral Commission, since there were no outside connections to the computer, and it was in a room that only our staff or Electoral Commission members could enter.

We tracked down the fraudulent vote totals, removed the illegal software, and corrected those entries in the computer before the final count. Our German women computer specialist slept by the computer each night thereafter. The revised system of faxed results plus phoned-in results from neutral observers allowed us to weed out the fraud.

To protect the process, we had set up two separate rooms for counting in the Electoral Commission Offices—one for phoned-in results by the observers, and one for faxed results by the head of each voting station. Students would put the two sets of results together matching the precinct sheets, and then Pirishaw and I would review the signatures and if everything was okay, we would certify the returns.[10] Matching the ballot faxed

[10] I was put in this senior position to help with the vote count because of my earlier work in South Africa with the ANC on their education strategy, my assistance with elections in six other African countries, and my work with twelve South African Historically Black Universities. Because this was the first truly national election in the history of South Africa, they had few Black specialists on elections and thus sought external help

results with those called in was tiring, exacting work for the students, Pirishaw, and me. The importance of the elections and cheerfulness and commitment of the students helped keep us on course. Plus, Pirishaw was a shrewd observer and open to everyone; thus, all the students were enamored of him.

In attempts to work around our careful precautions, several opposition operatives both faxed and called in fake results. We were able to identify which were the real votes by the signatures at the bottom of each result sheet which we could compare with our own list and the totals called in by official observers. This task was demanding, and all of us worked about eighteen hours a day, sleeping for four or five hours. Staff were in the counting rooms working 24 hours a day. We all were running on adrenalin recognizing that this was the final critical step after years of struggle for equality and democracy in South Africa.

The Electoral Commission had placed local police and internal security officials at both doors to our work area. Late on the second evening one of the young women returning from the bathroom noticed that security was gone from both doors. I checked the situation out with her; she was indeed correct. The absence of guards presented a real threat to security. Realizing that the opposition to the election might be planning to destroy or alter the result sheets, I immediately went outside to the army unit guarding us and protecting the building.

I said to the lieutenant in charge, a white dentist from Cape Town in the reserves as were all the soldiers, "We have a serious fraud problem and need your help. Could you assign soldiers to guard the two doors where we are working since the other guards have disappeared? I am extremely concerned that someone called off the police and the local security guards to create the opportunity to tamper with the election results and the data we have."

The lieutenant told me, "We are not allowed to enter the building; however, if any of my soldiers want to volunteer that is fine. I don't know anything about it." Like most people, the lieutenant wanted the elections to be fair no matter who won.

I said, "We need four, two at each door." Several men standing nearby immediately volunteered, and thankfully, we had our team. The soldiers stayed all night—in spite of a phone call from a general to the lieutenant

from people both the ANC and the Government thought would be fair. As a known commodity, I was thus an easy choice.

telling him to order the soldiers to leave the building. The lieutenant came up to our area to warn me and the guards that he and they faced punishment for disobeying orders. We wondered how the general knew we added the soldiers when the other guards left. None of the soldiers were fazed by the order and stayed until newly designated security returned—ordered by the Commission.

The soldiers said, "Fine, we are happy to run that risk in favor of assuring a free and fair election." Indeed, shortly after the soldiers were in place, a young white man was intercepted in the building with a hand grenade, headed our way. If he had been successful in tossing the grenade into the room where we archived the completed tally sheets, there would have been a national disaster!

In the end, thanks to tremendous teamwork and long hours of careful effort, the results represented the will of the people. The ANC won with Nelson Mandela becoming the first black South African head of state.

The magnitude of what we had accomplished did not hit me until several days later when the head of the Commission, a Supreme Court Judge, invited me for dinner at his home and thanked me for "saving the elections." I had lots of help in that, but it did show me how narrow the margin was between success and failure. We were also given great credit by some of the press. But most of all we were totally delighted with a job well done. When I ran into any of the students later in the year, I was given a big hug—much to the amazement of anyone who was with me.

On the night we finished counting, May 5th, several of us walked across the street from the Electoral Commission to the Carlton Hotel to hear Nelson Mandela's victory speech. The streets were overflowing with a multi-racial throng of thousands cheering, hugging each other, dancing, and singing. The tone and the joy provided a diametric contrast to the predictions of those who feared strife. This was a preview of the reconciliation that was to come.

Much has been written about Nelson Mandela, in my view the greatest political leader of this century. During the election, our work with him demonstrated the qualities that led to his success. He insisted that the elections be fair and stopped any efforts he was aware of on his side to use illegal means. He kept in touch with Pirishaw and me during the counting since the Commission (initially the two of us) in consultation with the winning party, were given the power to call the election once most of the ballots were in. At that moment, the ANC total was 67% of ballots counted—enough to change the constitution. Nelson Mandela did not want the ANC to be able to do that since he thought it would scare the White citizens whom he very much wanted to stay. He thought that enough anti-ANC ballots had not yet been reported and asked us our opinion on making the call.

We looked at the areas for which there were missing results and concluded that he was correct. He had explained that he did not want the ANC to have a two-thirds majority and be able to amend the constitution. That circumstance would allow the ANC to alter ownership of land which he feared would frighten the white population or lead to other laws to punish those in the former government. Pirishaw and I agreed that leaving counting open for another day would probably provide enough additional results from pro-government areas in the North to guarantee that the ANC would not receive more than 65% of the seats in Parliament. Mandela asked us to wait and call the election at that point. We were correct in our assessment and called the election when the ANC total was down to 65% of the vote.

March 17, 2008

Racism Revisited in South Africa

Having returned to South Africa in 2008 after a long absence, I was startled by the sharp contrast between the racial situation in the country then and when I first visited in 1992 and during the elections in 1994. Given the repression, violence, and humiliation perpetrated by apartheid, as well as the misinformation and myths fostered, the aftermath of the elections in 1994 with the calm that followed, the reconciliation, forgiveness, confidence, eagerness to move forward, and the apparent collective commitment to transform the nation was astounding. To be sure, a few white South Africans felt they could not live in a system of equality and left the country, a few others sought to set up a white-only settlement; however, those numbers were low. On the other hand, several groups of victims of apartheid demanded retribution, redistribution of white farms, and eschewed reconciliation. They too were a small but vocal minority which did not prevail.

Post-Election South Africa

I had been working with twelve historically black universities and technikons since 1993—a project which continued until 1997. After the elections, I was struck by the fact that even supporters of apartheid among the university and technikon faculty members, staff, and students, seemed relieved. Many who had supported apartheid apologized for not understanding the depth and violence of racism, others for believing government propaganda about the ANC, or regretting their doubts about the future. During the struggle period, race did not define who opposed apartheid, and race did not define my experience as a white outsider. Many of the strongest proponents of equality were white. The question was not one of race, rather of a person's beliefs, how he/she treated people, and one's behavior in a context of repression and violence against the black population. Following the elections, the majority population, white and black demonstrated a phenomenal graciousness toward those who had not joined the struggle, a willingness to forgive and move on. This willingness to forgive amazed me and many South Africans.

The Truth and Reconciliation Commission (TRC) was created to promote reconciliation by allowing those guilty of racial crimes (on both sides) to confess their actions and show genuine remorse. If those guilty convinced the Commission that they were contrite and revealed all the illegal acts they had carried out, they could avoid prison.

The TRC was led brilliantly by Bishop Tutu. Hearings were broadcast live, and I listened to them while driving between historically black institutions from one end of South Africa to the other. The process of reliving those horrors was difficult for the victims who testified; the thought, the care, and the gentleness of the Truth and Recreation Commission process promoted reconciliation and allowed many people to finally learn the truth of the fates of their loved ones and others to put their fear, loss, and anger behind them. Surprisingly, most of those accused of crimes fully admitted in detail the actions they committed including terrible examples of torture, murder, and other violence. However, not all of them passed the sincerity test and some were sent to jail.

The Commission hearings produced the most remarkable period of awakening, new understandings, generosity, and civility I have ever heard of or witnessed anywhere in the world. This was a period of remarkable transformation for many whites who had supported the apartheid government. Listening to the Commission hearings and at the same time working at a number of historically Black universities and technikons gave me a window into people's feelings. I was amazed at the sincerity of forgiveness and the openness of several who had been supporters of apartheid on campuses.

Much of the change in attitudes I witnessed occurred on university and technikon campuses—at Fort Hare, Peninsula Technikon, University of the North, Pretoria, Cape Town, University of the Western Cape, University of Natal, M.L. Sultan Technikon, and others. I do not want to overly romanticize this period; there were detractors, efforts to create white racist redoubts in several parts of South Africa, demands to take the land of white citizens, and to subvert the progress made at several universities.

Nelson Mandela influenced the success as he worked tirelessly to make nervous Whites feel secure in part by ignoring the white separatists, helping integrate the national soccer team, singing both the African and Afrikaner national anthems at the post results celebration at the Carlton Hotel across from the Electoral Commission, having two tall white policemen as his bodyguards, and welcoming everyone.

To be sure, not all people, white or black, embraced the new order of reconciliation. Nonetheless, the level of reconciliation and the collective efforts of all races to build a new South Africa were extraordinary and visible as demonstrated in attitudes in shops and restaurants, open mixed racial dating, music concerts, and theaters without separate seating. Some racial interaction had occurred quietly or in secret in the past—now those actions could be out in the open, including interracial dating.

The South African renaissance taught me a powerful lesson, coming from the US, which has still not dealt fully with its history of slavery and racism and the continuing institutional racism. Though there has been progress in the US, racism remains a constant problem with incidents at our universities and elsewhere—cross burnings, nooses hung to intimidate, racism in workplaces. Racism is still a constant and major struggle in the US. The young in the US do not remember the civil rights marches, Martin Luther King's "I Have a Dream," or the horrors of slavery, lynching, mass killings, and the ravages of racism. A number of US schools, colleges, and universities have learned that multi-culturalism must be retaught; that avoiding racism is a constant battle that requires ongoing education programs and reinforcement of norms of equality, fairness, and justice, and strong overt action when racism appears. There has also grown up a right-wing group that seeks to outlaw teaching about institutional racism in the US, a major reversal of progress made to date in equality. The "Black Lives Matter Movement" is having positive effects in several cities and states, but there is a long way to go to achieve equality.

Police must be retrained and overseen since much of their past experience has been to reinforce racism and mistreat people of color. Most importantly, institutional racism in health care, housing, employment, agriculture, and education must be ended.

February 2008 Return to South Africa

Upon returning to South Africa in February 2008, I immediately realized that lessons learned the hard way in the mid-to-late1990s seemed to have been forgotten by many people. Sadly, especially from 2007, there began to be a reversal of a number of the positive changes toward equality. During the Zuma presidency, especially, a tremendous amount of corruption by the ANC and a number of businesses brought out the worst fears of Black and White South Africans. Much of the civility and graciousness and the lessons of forgiveness and openness were lost. In their place, a new racism emerged. This seemed particularly the case among the young who had not experienced the earlier struggles or the example of graciousness of Nelson Mandela, Desmond Tutu, and others. Notions of superiority long suppressed were reemerging; a sense of entitlement was evident once again among parts of the White population and taken on by some of the Black population as well.

During this trip, I witnessed racism toward black students in Cape Town. Press coverage of incidents at the University of the Free State against Black staff and other demonstrations of racism were reported in its wake. While most people took these seriously, others viewed these incidents as "… merely a joke," "…boys will be boys," "…this has been taken out of context," "…these young men are being used as scapegoats."

I think not. The young white men knew they were trying to maintain white elitism in their residence halls, they intended to humiliate and demean Black workers, and they were trying to assert white dominance. Yet in that situation, as Vice Chancellor Jonathan Jansen astutely suggested,[11] South Africa was in need of a new engagement—renewed efforts of the sort that fostered equality over a decade and one half ago.

Part of the problem was that some aspects of apartheid had been institutionalized. The laws could be changed to ensure equality in employment and salaries, for example, but how they were carried out was another

[11] Personal communication March 2008.

matter—especially in areas in which there was some flexibility available to those in power such as salary level or position. Nonetheless, the struggle for equality goes on with major voices working to make it work. In many ways, the realization of the problem is clearer in South Africa than in the US which has had many more years to solve it. I remain hopeful for continued success in South Africa after backsliding under the corrupt President Zuma, by the recently elected honest, hardworking President Cyril Ramaphosa—though his task remains tremendous given the corruption in the ANC and the economic decline that has resulted. Nonetheless, progress is being made.

The transformation of South Africa is an excellent example of what mass public demands and support at every level can accomplish even in the face of draconian legislation, massive police and military violence, and suppression of Black communications and free speech. While the South African struggle for equality had remarkable leadership in Nelson Mandela and a number of other outstanding leaders, it would not have happened without the activity of local leaders and the public—both Black and white. The struggle was a long one with many people killed in the process, yet even the die-hard leaders of apartheid South Africa eventually realized they could no longer govern by suppression and force. People at every level were working to end apartheid and there was no way to stop it—try though they did. I can think of no other comparable example of the success of broad citizen action and solidarity anywhere in the world that equals that in South Africa during this period. We see reflections of this in Afghanistan, but in that case the major efforts and successes were in higher education—changes that could not have happened in the country as a whole at that time, as we will see. Nonetheless, both cases demonstrate the critical importance of broad citizen support and activity at many levels along with outstanding leadership.

CHAPTER 5

Pakistan: Top-Down Transformation

Introduction

Pakistan is particularly interesting for several reasons. The seventy-four-year-old independent country, along with neighboring India, received its independence from Great Britain at the same time in 1947.[1] India focused on massive improvement of higher education from that time becoming one of the best systems in the region with some of the top higher education institutions in the world. Pakistan did little except adding a small number of student enrollments. As The World Bank said of that period in 2006:

> Decades of neglect have drawn universities in Pakistan – and more generally the higher education sector (HESS) – to levels which are incompatible with the ambitions of the country to develop as a modern society and a

[1] Pakistan, officially called the Islamic Republic of Pakistan, is located in South Asia, surrounded by the Arabian sea and the Gulf of Oman. It is bordered by India, Afghanistan, China and Iran. It has a population in 2021 of about 225 million. Prior to colonization by the British, it was the home of many of the most ancient civilizations in the world (Wikipedia 2021), "Islamic Republic of Pakistan," downloaded on July 28, 2021, from: https://en.wikipedia.org/wik:/Pakistan.

F. M. Hayward, *Fostering Institutional Development and Vital Change in Africa and Asia*,
https://doi.org/10.1007/978-3-031-04364-2_5

> competitive economy. As it stands now, the subsector does not compare well with its counterparts in the region, and unless profoundly reformed, it may become an obstacle to the continuation of the rapid economic grown, instead of becoming its main engine.[2]

The Report went on to say that what "plagues" Pakistan is that it is both "small in size and low in performance." It continued: "Today [in 2002] these elements are in place with the creation in 2002 of the Higher Education Commission (HEC)."[3] It was not until 2002 that serious changes were planned to improve higher education and catch up with the rest of the world through preparation of a Medium-Term Development Framework (MTDF). In spite of several excellent reviews of the needs of higher education over the years and proposals for change between those years, nothing happened. Another striking aspect of the change effort in Pakistan is that unlike the other five examples explored in this manuscript, the plans in Pakistan were unabashedly top-down, though that was to have negative consequences several years later. The results of these top-down efforts were outstanding leading to substantial transformation taking place. This was facilitated by strong support from President Musharraf and an excellent strategic plan (MTDF) prepared by the Higher Education Commission (HEC) led by Professor Atta-ur-Rahman, and written by Sohail Naqvi, Deputy Director, and Kamran Naim, Project Coordinator.[4]

India by contrast had seven universities among the world's best[5] while Pakistan had none. At the same time, primary and secondary education in Pakistan deteriorated in quality making too many graduates of secondary school below par and lowering the quality of many of those admitted to higher education. After five years of excellent quality improvement with substantial funding from both the Pakistan government and The World Bank, higher education's quality declined for the next decade. Much of

[2] World Bank. (2006), "Higher Education Policy Note Pakistan: An Assessment of the Medium-Term Development Framework," Human Development Sector, South Asia Region, note 1.

[3] Ibid., note 2.

[4] Interview with Kamran Naim on March 14, 2008.

[5] Mint. (June 23, 2021), "India Has Seven Varsities among the World's 200 Best Young Universities," Live Mint, downloaded on June 30, 2021, from: https://www.livemint.com/education/news/india-has-seven-varsities-among-world-s-200-best-young-universities-11624449551128.html.

this was due to lack of continued government support as well failure to inform and mobilize the public generally about the needs of higher education. At the time of writing in 2021, Pakistan's education had still not recovered from this decline providing many examples of actions and disasters that can lead to the long-term failure and decline of higher education. The irony is that the phenomenal improvements from 2002 to 2008 were broadly evident yet did not slow the attacks on the HEC or education starting in 2008 for many years thereafter as we will see in what follows.

Transformation of Higher Education Starting in 2002

The transformation of higher education begun by the Higher Education Commission (HEC) in 2002 and built on earlier work by Professor Atta-ur-Rahman when he was Minister for Science and Technology. When he was appointed head of the HEC in 2002 by President Musharraf, he immediately pushed to upgrade and expand IT, established the Digital Library with access for all public and private higher education institutions, and a Pakistan Research Network to encourage and expand faculty research.

All this took place in the beautiful capital, Islamabad, build in the 1960s as a planned capital to replace government offices headquartered in Karachi in the 1960s, with wonderful creative architect-designed buildings, wide boulevards, parks, and exquisite shops. There were excellent hotels, wonderful restaurants, and unostentatious security. It was one of the most pleasant places to work in my fifty years of experience.

The initial focus of new money from Government was on faculty development. The HEC sent about 1000 faculty a year abroad for Ph.D. training for several years. Each could apply for a $100,000 research grant when they returned with their Ph.D.s. The HEC also made sure new Ph.D.s had a job when they returned as assistant professors. Emphasis was on the sciences, and while these students were away, the HEC upgraded labs and bought state-of-the-art sophisticated equipment for them. They also set up a number of large labs in several regions available to all faculty members.[6] The HEC had a remarkable 97.5% return rate for those sent abroad.

[6] Interview with Atta-ur-Rahman on August 4, 2002.

During 2004, the HEC worked on plans for implementation of the Medium-Term Development Framework (MTDF). Its major foci were: (1) the quality of higher education; (2) improving access for students from a low of 2.6% of 17- to 23-year-old students in 2004, to 12% by 2015; and (3) fostering greater relevance of higher education to the nation while reaching international norms.[7] There had been a long discussion within the HEC about its focus before they settled on the three priorities. Faculty development, especially Ph.D.s, was among the first actions taken with 3820 students sent to Ph.D. programs, more than 1000 for foreign degrees, right away.

A four-page outline for the MTDF was sent to the Planning Commission, which was enthusiastically approved by them. The HEC then proceeded to flesh out the MTDF. Following the focus on faculty development, a tenure-track system was added (on a voluntary basis) with significantly higher salaries for those who qualified. That produced a substantial improvement in both the quality of teaching and the amount of research, the latter growing from about 800 publications a year in 2001 to more than 12,000 publications in 2015.[8]

When the MTDF was completed, a team from the World Bank arrived to join the HEC team in planning implementation of the project.[9] This was one of the processes especially important about The World Bank funding plans. Unlike some other funders which do not participate until all the planning is completed, once the Bank decides to support a project, they become involved in its planning as well as its funding. That approach improved the quality of implementation and saved a great deal of time on both sides. That consultation by the joint World Bank and HEC team included a careful assessment of costs and the income expected to be raised to meet the HEC Plan goals. The teams went through a wide range of both cost and income scenarios and adjusted the plan to meet those realities once both sides were satisfied with the projections. Careful backup plans were prepared for adjusting the MTDF if funding was below expectations—areas to cut, places in which its scope could be reduced,

[7] Ibid.

[8] Ibid.

[9] The team was led by Benoît Millot and was comprised of Mohammed Allak, Fred Hayward, Norman LaRocque, Yoko Nagashima, Naveed Hassan Naqvi, and Jamil Salmi.

and potential additional sources of income if needed. Substantial funding was promised by both the Government and The World Bank.

The Bank team looked at potential spending patterns in higher education for Pakistan over a fifteen-year period, carrying out a low conservative estimate and a high projection. These assessed the costs of the MTDF programs, their feasibility, comparisons of the costs of the programs and the projected income looking for gaps between the two. The Bank projections were based on costs, enrollments, and resources.[10] Among the most difficult to assess were possible enrollment numbers over time. Even at that time, at the outset, the HEC had three different sets of enrollment numbers depending on which unit you asked. This was not an encouraging finding at the outset. However, with a great deal of time, effort, and digging the teams reached an accord on the most accurate data and what seemed to be the best backup plans.

The World Bank's role in the process of budget and implementation planning for Pakistan was critical in that it brought in a wide range of academic and economic expertise to the already well-trained HEC staff—people with broad experience in building high-quality education systems around the world, and thus the ability to provide ideas and depth to the implementation planning. For example, planning about accreditation, which is a complex issue in any country, was helped by suggestions and examples from the experiences of World Bank staff members who had worked in a dozen countries around the world on that process observing both excellent successes, as in South Africa, and terrible failures in places such as Nigeria at that time.

Especially important to success was the outstanding leadership and staff of the HEC. Professor Atta-ur-Rahman, the director of the HEC was a true visionary and had the ability to bring on board the President of Pakistan, legislators, and most higher education leaders with his eloquence and brilliance. He was a noted organic chemist with many international awards for his research and writing. Among the most famous is his Series: *Studies in National Products Chemistry*. The Bank team found him engaging and helpful though some of his colleagues saw him as intimidating. That was blunted by his excellent staff.

Dr. Sohil Naqvi, the Deputy Director, was one of the most astute financial experts I have worked with in my higher education career, constantly

[10] World Bank (2006), ibid., p. 66.

checking budget projections on his computer (or in his head), suggesting solutions to problems, and actively leading the Pakistan team in the negotiations. He was also a warm and supportive leader who was universally liked and respected by the staff. He was a handsome figure, and always open to people giving those who worked with him a sense of confidence in both his and their work. He was an excellent partner to Atta-ur-Rahman and the two worked together easily with great respect for each other.

In addition, a number of other outstanding mid-level staff members including Kamran Naim, and a number of women, such as Zia Batool, were driving forces behind the successes in accreditation and quality improvement and in other areas. Kamran Naim, in addition to being one of the two main authors of the MTDF, was also largely responsible for setting up the Digital Library for Pakistan's higher education. He managed to find several foundations abroad that subsidized access to journals or provided them free or at low cost to underdeveloped countries. Pakistan was able to access more than one million dollars worth of journals for a small fraction of that price. Several publishers also provided journals and books free or at cost to underdeveloped countries. Kamran Naim took advantage of all of these opportunities to make the Pakistan Digital Library one of the best in Asia. A few years later he also helped Madagascar set up a Digital Library. Being even poorer than Pakistan, he was able to arrange for them to have the same access as Pakistan for nothing. It was an amazing gift to Madagascar and made a major difference in upgrading those facilities.

Kamran was an outgoing person in his own right at HEP, respected and liked by staff. Yet at the same time he was a very private person, studious, careful, and thoughtful—a planner with great foresight. He had great support from his family, who periodically invited several Bank staff over for dinner. His father was retired but clearly on top of all the political issues and intrigue going on at the time and Kamran had a highly successful sister who adored him. His father, as a hobby, had become a specialist in Persian carpets, from which some of us benefited both as to price and quality of our own purchases. It was clear that Kamran gained many of his superb leadership and writing skills from his father, as well as a way of carefully examining all issues.

Another example of the outstanding work of HEC staff members was demonstrated by Zia Batool, who led the quality assurance effort, largely on her own under the minimal guidance of Dr. Riaz Qureshi who headed four divisions including Quality Assurance, with help from several senior leaders. Zia Batool organized the quality assurance process from scratch including plans for setting up what were called Quality Enhancement Cells (QAC) at each institution. That process was carried out in a very consultative manner with initial conversations within the HEC to develop the standards, followed by consultation with the Quality Assurance Council, and then discussion with individual universities to put the process in place. The implementation itself was hampered by the fact that the HEC had very limited staff dealing with Quality Assurance, but it was well underway in 2003 and 2004 with several institutions having completed their self-assessments. Most institutions were surprisingly supportive, unlike some universities in several other countries, recognizing that accreditation would be a good sales pitch to the public for their institutional recruitment. Examples of the standards put in place are listed below.

Criteria and Standards

1. Program Mission, Objectives and Program Mission, Objectives and Outcomes (3 standards).
2. Curriculum Design and Organization (7 standards).
3. Laboratories and Computing Facilities (2 standards).
4. Student Support and Guidance (3 standards).
5. Process Control (5 standards).
6. Faculty (2 standards).
7. Institutional Facilities (2 standards).
8. Institutional Support (3 standards).[11]

I attended several meetings with Zia Batool and university leaders. She had a gentle but persuasive way of making the case for accreditation and helping institutions set up the *quality assurance cells* which were required of each university. While Pakistan as a whole had a long way to go in reaching gender equity, substantial progress had been made in higher education including priority for hiring women faculty members. Sexism was never apparent in the relations of Zia Batool with HEC staff or with the universities. Several universities had made substantial progress

[11] From presentation by Zia Batool, personal notes.

in recruiting women faculty members and promoting some of them to leadership positions. It was clear to us, that once an institution reached about one-third of women, sexism stopped.

A shortage of staff and the resulting problems limited the ability of the HEC to work with all institutions on the new accreditation policies, including to visit them to help start the accreditation process. That was to foretell later ongoing problems for the HEC. Funds being in short supply, 74 of 84 projects to be carried out by the Quality Assurance Agency (QAA) by September 2008 were cancelled because of lack of funds.

Kamran Naim and Zia Batool were part of the secret of success of the transformations that took place in higher education in Pakistan from 2002 to 2008 in implementing the MTDF. They are another example of the importance of middle-level people to successful change. Both were responsible for major achievements in the implementation of the MTDF.

Universities were eager to obtain the grants to send faculty members for Ph.D. study at home and abroad so the program to upgrade faculty training was an immediate success. Getting accreditation underway was more difficult but with help from the Bank team made excellent progress under the leadership of Zia Batool. Zia was very generous, regularly providing snacks for the office staff, wanting to pay if team members went out to lunch, and was a lovely person in all respects. She was the first wife of a multi-wife family and devoted as much time as possible to their children. She traveled to Afghanistan to help the Ministry of Higher Education there get accreditation underway. Even when the HEC was under brutal attacks, she worked hard to keep accreditation going.

A number of other successes occurred during this early period of implementation of the strategic plan, including the addition of distance education which became very popular. Enrollments grew in both public and private institutions from 183,000 in 2003–2004 to 316,278 in 2006–2007, an increase of 72%. The HEC's development funding grew during this period by 33% which profoundly effected the improvement of facilities. The Digital Library provided access to over 20,000 research journals, e-books, and other materials for faculty members, staff, and students. This digital library later became a model for several other countries. Funding increases, including for research, faculty travel, and conferences increased by 340% from 2001/2 to 2005/6. Yet this came after years of underfunding and still left Pakistan behind half of other higher education systems in the region. Per student funding rose 41% over this period. To some extent, the government brought funding levels up to what they should have been in the first place to enable higher education to upgrade capacity to proper levels.[12]

The HEC was also able to improve the physical and technical infrastructure of higher education institutions. The Commission worked hard to ensure merit-based appointments and promotions reviewing many institutions to guarantee quality. Government also wanted to encourage private provision of higher education recognizing that demand in Pakistan was far beyond the capacity of government to meet it. During this period, the number of higher education institutions increased from 44 in 1997 to 122 in 2008, an increase of 186%. Overall, this period marked the impressive implementation of MTDF over five years, though it started from an exceptionally low base given the lack of government interest in higher education between 1947 and 2001. Part of the success was due to the dogged effort of the HEC to improve quality and meet or better its institutions in relation to international standards.

However, by 2008, the HEC was under major attack by members of Parliament and some regional governments which wanted to wrest control of higher education from the HEC. As one Vice Chancellor said at the time, "the balloon of the HEC is burst now."[13] They felt that they no longer needed to do what the HEC suggested. This marked the beginning of the weakening of the HEC.

The attacks on the HEC emanated from several fronts. Some of the hostility was based on anger in other Ministries at the favored status of the HEC (which was a quasi-ministry) in terms of funding thanks to the strong support and prioritization of President Musharraf. Several regional governments were unhappy with the HEC for its monopoly of power over higher education and wanted that shifted to individual regions. The crowning blow, however, came as a result of a Supreme Court order to the HEC to review the qualifications of members of Parliament, whom by law were to have a higher education degree. The review by the HEC demonstrated that about 200 MPs, as well as the Minister of Education, had bogus degrees. A great deal of pressure was then put on the HEC to say these degrees were legitimate.[14] The HEC refused to comply, but

[12] World Bank. (2006), "Higher Education Policy Note: An Assessment of the Medium-Term Development Framework," World Bank, Washington, DC.

[13] This section is based largely on an interview with Zia Batool on September 2, 2008.

[14] Ibid. Atta-ur-Rahman.

the remaining MPs were outraged and turned on the HEC. Thus, the Commission was under combined pressure from Parliament, a number of ministries including finance, and several regions including Find and Punjab which wanted control over their own universities.

Parliament tried to abolish the HEC during this period but was prevented from doing so by the eighteenth amendment to the Constitution and a Supreme Court decision upholding the existence of the Commission. At this point President Musharraf had left office, and the People's Party had come to power. They were hostile to the HEC and early in August 2008 terminated funding for several thousand scholars studying abroad.[15] This was a disaster for the students abroad, most of whom were from poor families, as well as for their universities. At this point, Dr. Atta-ur-Rahman, well aware of the opposition to his leadership in the new government, partly because of his close association with the former president, offered to resign if the funding was restored, although he still had two years on his contract and could not be fired except for cause. The government agreed, the funding was restored. That did not, however, stop attacks on the HEC, now led by Sohail Naqvi who then became the focus of attacks. He was forced to resign in 2010.

The higher education system was now divided into two parts, with some institutions under the control of the provinces and some under the HEC. This further weakened higher education and fostered a downward spiral in quality—in part because of a series of major funding cuts by the government; in part due to limited funding in the affected provinces; and because of the division of authority. While a new government was elected three years later and restored much of the funding, the damage had already been done and higher education was once again suffering from poor quality and continued to be far below international standards.

Major Challenges to Higher Education

At this point serious challenges threatened higher education. Perhaps the most serious was maintaining the emphasis on quality and merit hiring. With the division of authority, those goals became difficult. The regions were not equipped to evaluate quality and failed to do so in some cases, leaving quality assurance solely up to institutions themselves. Some did

[15] Ibid. Atta-ur-Rahman.

an excellent job, but others ignored the issues—at their peril. Then there was the problem of gender equity for staff and students. While substantial progress was made during the early HEC efforts, gender equity was no longer a priority for many institutions. Nepotism and favoritism reigned. Then there was the problem of reduced funding both by the government and the regions. While student numbers continued to grow, staff expansion did not keep up, nor did continued expansion of desperately needed laboratories, other facilities, and supplies.

Pakistan's problems were compounded by lack of unity in the HEC at that time and in higher education in general. Thus, once again, there was no agreement on how to solve the growing number of problems or deal with crises when they existed. There was also a decline in the effectiveness of the leadership in some of the universities now with less oversight. All these problems were compounded by the increasingly poor quality of primary and secondary institutions, which were mostly under the control of the Ministry of Education, which was also suffering from lack of quality leadership. The HEC and the universities had tried to stem this decline to the extent they could from 2004 to 2010 but given their reduced budgets and staffing problems they were no longer able to provide significant assistance. This had the additional effect of lowering the caliber of many of the new students admitted to higher education. Added to that, the top-down transformation successes of the HEC from 2002 to 2010 meant that there was little public knowledge or support for the HEC. The cumulative effect of these problems helped accelerate the decline of higher education generally.

Current Situation

Thus, in spite of the major progress from 2002 to 2010, the situation has deteriorated since then. No Pakistan University is in top 500 world-wide and only 17 are in the top 1000, unlike India which had several in the top 200 world-wide. This is a sad commentary on the positive direction of Pakistan's higher education starting in 2002. Indeed, higher education rankings declined by 20% between 2019 and 2020.[16] Added to

[16] Raza, Ahsan. (September 2, 2020), "No Pakistani University in top 500 Times Higher Education World Rankings 2021," downloaded on July 5, 2021, from: https://en.dailypakistan.com.pk/02-Sep-2000/no-pakistani=university-in-top-500-times-higher-education-world-rankings-2021.

that is the huge brain drain of outstanding faculty and graduates fleeing a deteriorating system to take more promising jobs—meaning that the country was losing many of the best and brightest of its people.

The earlier efforts at gender equity seem to have ground to a halt. Women in general are getting an inferior education to men in primary and secondary school, which puts them at a disadvantage in gaining admission to universities. Most who do go on to college end up in lower quality institutions. The quality of primary and secondary education is also significantly affected by wealth, with those areas of high-income citizens having much better schools than poor areas. This continues to exacerbate the economic inequalities of the country. Roughly 70% of women work outdoors—most in gardens and fields—which also limits opportunities for young girls to go to school. Women are expected to work in the home, have and raise children, and do what their husbands tell them. This is little changed over the last decade.[17]

Pakistan continues to be a patrilineal society. Education for boys is prioritized over girls. Female literacy is only 30% with Pakistan ranked 130th in human development partly because of low levels of spending on education.[18] Pakistan was ranked 92nd of a low of 94 in gender equity in 2010. While the overall rate of education was 58.3% in urban areas, literacy was only 28% in rural areas and 12% for women in those areas. Over 2000 Islamic schools for girls exist, but they are of mixed quality. More than 400 girls' schools have been destroyed by the Taliban, who opposed education for females.[19]

Education levels for boys are much higher than for girls. Although public primary level education is compulsory for all children Pakistan has not achieved that goal with only 79% of students enrolled. By middle school, literacy is 62% of students with 54% boys and 46% girls.[20] Interestingly, however, by higher education, the number of women almost equals that of men. A study by Monazza Aslam concluded the investment in

[17] Wikipedia. (2021), "Women's Education in Pakistan," downloaded on June 25, 2021, from: https://en.wikipedia.org/wik/Women%27s_education_in_Pakistan.

[18] Ibid., p. 2.

[19] Ibid., p. 4.

[20] Ibid., p. 6.

women's education had a higher return than that in men's higher education.[21] Unfortunately those findings have not affected policy. Women are also higher achievers in higher education than men in university education. Sadly, most Pakistani women do not get that chance.[22]

Implications of the Experience in Pakistan

Pakistan had great success in improving higher education from 2002 to 2010 with an outstanding strategic plan, quality assurance, upgrading the sciences, increasing the number of faculty members with Ph.D.s, including sending more than 1000 abroad. The HEC also expanded facilities, in particular with increases in the number and quality of science labs including several extensive state-of-the-art labs shared by several universities.

The transformation of higher education in Pakistan was self-consciously top-down led by Dr. Atta ur Rahman and Dr. Sohail Naqvi, with extraordinarily strong support from President Musharraf. Professor Atta-ur-Rahman emphasized the top-down approach feeling that it was the only way to move Pakistan's higher education from its lack of significant progress since independence in 1947. There was some opposition from several universities but at the outset, as noted earlier, most institutions were so happy with the emphasis on, and funding for, faculty Ph.D. training, laboratories, and other upgrades that they did not complain. In spite of his many superb talents, Professor Atta-ur-Rahman was not particularly good at collaborating with university leaders, faculty members, or students. Whether he was correct in his assumption that progress on transformation could only be made from a top-down approach remains open for discussion. However, in the long run he needed to gain broad higher education institutional support, that of the public, political parties, and Parliament. In many parts of the world, education—especially higher education—is a non-partisan issue. This could have been the case in Pakistan with a concerted effort by the HEC to fully involve, or at least inform, higher education institutions in the process, as well as members of the public and politicians. However, the HEC did not focus on such

[21] Ibid. As quoted in Wikipedia (above), p. 4.

[22] Kahn, S. (2007), "Gender Issues in Higher Education in Pakistan," *Bulletin*, Issue 162, 2007.

an effort or lay the groundwork for strong public support of the kind that took place in South Africa, Ghana, and Afghanistan. This meant that when the new government in 2010 decided to attack the HEC and reduce funding, few supporters mobilized to oppose those efforts.

President Musharraf made substantial increases in funding for higher education which had been underfunded since 1947—one of the worst in the region and way below neighboring India which had remarkable success with 12 higher education institutions in the top 100 in the world.[23] Higher education grew in number of institutions and in students in Pakistan during the period from 1947 at which time it had only one university—the University of Punjab—from 2002. However, by 1997 Pakistan was in disarray with 35 Universities of which only 3 were federally administered and 22 were under Provincial governments. The total enrollment at that time was almost 72,000.[24]

In many respects this case emphasizes the concern about the need for scholars to focus both on top leadership and on staff, faculty members, administrators, students, and others at lower levels to understand the obstacles to change. In the case of Pakistan, in contrast to South Africa, Ghana, Afghanistan, and Madagascar, no major attempt was made to gain broad support—and even fostering involvement in planning by higher education institutions was limited. They were happy to have the funding for science laboratories and money to send faculty for Ph.D.s both locally and abroad. But they were not integrated into the change process and when those changes were attacked, few people made the case against these attacks.

Professor Atta-ur-Rahman was forced to resign in 2008, followed by Dr. Sohail Naqvi in 2010 after major assaults on the HEC. Pakistan's higher education has not recovered its former quality since then. In 2016 Atta-ur-Rahman thought, "The momentum we set up at the time was irreversible. Because thousands of these students came back, so there was a steamroller moving in spite of attempts by the previous government to

[23] Yahoo News. (March 4, 2021), "12 Indian Institutes in Top 100 of QS World University Ranking; Pokhriyal Credits Govt's Education Reforms," downloaded in July 2021, from: https://in.news.yahoo.com12-indian-institutes-top-100-16245124.html.

[24] State University. (n.d., circa 2020), "Hybrid Learning Flexible Online Option," downloaded on June 26, 2021, from: https://education-state.university.com/pages/1146/Pakistan-HIGHER-EDUCATION.html.

disrupt the system, so things are now back on track."[25] The professor was wrong and higher education in Pakistan has, for the most part, fallen to the mediocre status of the pre-2002 period.

In 2020, Hazir Ullah wrote a piece entitled, *Quality in Higher Education: What is Wrong?* in which he lamented the decline in the quality of higher education in Pakistan over the last decades, noting the HEC's efforts to reverse that trend, but suggesting that not enough was being done. He stressed: "The declining standard of education is a complex and complicated problem and may not be redressed by revision of curriculum." He went on to say, "Embedding quality in higher education demands all stakeholders (HEC, universities' administrators, teachers, and students) to join hands and put our heads and resources together to achieve it."[26]

In its report on the Medium-Term Development Framework in 2006, The World Bank recognized the need of the HEC to build public support, "The reforms launched by the HEC, and those which are in the MTDF pipeline are meant to radically change the academic world in Pakistan. They are met with enthusiasm by some and by resistance by others. They are not irreversible yet. To succeed, the HEC must establish its legitimacy and improve its public image. Above all, the Commission must invest in open consultation and communication with the entire community it aims to serve."[27] Sadly, those things did not happen.

The critical point for long-term higher education improvement in Pakistan is that although the transformation was successful as long as a particular President was in office and the main leaders of change in higher education continued to lead the HEC. The lack of broad support even in higher education, and no major effort to gain public support, led to its later problems including underfunding by governments. Complicating the situation is that lack of funding has led to increasing student numbers to help balance university budgets but resulted in larger classes with lowered quality in many cases. The contrast of the Pakistan situation

[25] Interview with Dr. Atta-us-Rahman on August 4, 2016, by the author.

[26] Ullah, Hazir. (September 30, 2020), "Quality in Higher Education: What Is Wrong?" in *The Nation*, downloaded on June 26, 2021, from: https://nation.com.pk/Reporter/dr-hazir-ullah.

[27] World Bank. (June 28, 2006), "Higher Education Policy Note. Pakistan: An Assessment of the Medium-Term Development Framework," Human Development Sector, South Asia Region, Report No. 37247.

with the situation in four of the five other countries examined including Afghanistan, South Africa, and Ghana (the biggest exception being Sierra Leone) demonstrates the critical nature of broad support at every level to sustain change and transformation. At the time of writing in 2021, the situation in Pakistan did not seem primed to change for the better any time soon.

CHAPTER 6

Building Higher Education Amidst War in Afghanistan

The country was destroyed by people who were experts in destruction. It will take experts of even greater ability to rebuild it.[1]

"Damn!" we heard from the cockpit as the Boeing 707 with about 200 people aboard neared Kabul in June 2003. The pilot then lit the seatbelt sign and came on the air to sternly warn us to tighten our seat belts. Taliban activity in the hills around the city necessitated a rapid landing, circling down abruptly from directly above the airport rather than the usual slow descent through the mountain passes. When he introduced himself to the passengers after we boarded the aircraft in Dubai, he told us he was a former fighter pilot with more than ten years' experience—a statement that later assuaged my fears before our frightening landing.

Suddenly, the pilot dropped one wing and the plane fell 25,000 feet in dizzying tight spirals landing perfectly on the runway and roaring to a stop. My stomach did not catch up with the rest of me for an hour or more. Seated by a window, I could witness the detritus of war in all the wrecked airplanes lining the runway including destroyed Russian jets hauled to the perimeter fence. Many planes had been shot down by the

[1] Mr. Hilalluddin Hilal, Deputy Minister for Security and Police in a meeting with the author and Sara Amiryar, June 12, 2003.

F. M. Hayward, *Fostering Institutional Development and Vital Change in Africa and Asia*,
https://doi.org/10.1007/978-3-031-04364-2_6

straight-shooting pilots of the Mujahideen and Taliban with their own MiGs in the fighting near the end of the Russian occupation.

The Russians wanted these planes out of sight because they were reminders of their having dropped thousands of personnel land mines in most rural areas of Afghanistan. The Russian goal was to drive Afghans out of the agricultural areas into the cities where the Russians thought they could control them. They succeeded in making most of the country's agricultural land unusable, though they never were able to control the Afghans even though much of the rural population was forced to move to the cities. Unfortunately, these "bouncing mines" as they were called, (if you stepped on one, the mine jumped up about five feet and then exploded maximizing the death and injury) were everywhere. They were still killing almost 50 Afghans a month five years later—down from 200 a month at its heights—mostly women and children gathering firewood. Demining had been underway for more than a year, led by the Norwegians and the United Nations. This was a massive and dangerous job still done primarily by hand since no alternative worked. In spite of bullet proof vests, several deminers lost limbs in the process. This was my sudden "welcome" to Kabul and to life in an active war zone. I am not easily scared, yet I was intimidated by this entry and relieved to be on the ground alive. I was to lead a team to assess the possibility for private higher education—then illegal in Afghanistan.

June 10, 2003

Kabul, First Impressions

People were especially helpful, and my Afghan/American colleague Sara Amiryar continued to amaze me with the amount of work she accomplished as well as the value of her contacts. For example, the stores were out of cell phones or telephone numbers everywhere in the country—yet I will get a phone tomorrow thanks to Sara. I do not know how she does it, and I never asked. I am delighted with our progress already. We will start focus groups here in Kabul next week and go to several towns in outside areas such as Herat—those that are deemed safe, and there were quite a few at this point.

Kabul is an amazing city. Once I started looking carefully, I noticed pockets of rebuilding. For example, two houses down from our office there had been a brick and concrete three-story house that was totally bombed out. The debris, which were untouched when I first arrived, were later being cleared out by hand, simply using hands, sledgehammers, and shovels, and the beginnings of a new foundation was being put in. A good deal of business potential and energy was evident in Kabul with a strong desire to start over again—even after 25 years of war! At the same time, the problems were major: most people had lost nearly everything and, thus, had little if any money to start reconstruction. Finding building materials such as cement and steel reinforcing rods was difficult, and little remained of usable construction equipment. Nonetheless, the changes in Kabul in the four weeks we had been there were heartening.

December 2005

Working in the Ministry of Higher Education

Two years later, at the recommendation of the World Bank, I was hired in 2006–2007 through the University of Massachusetts, Amherst, to return to Afghanistan because of my years of work on strategic planning and accreditation around the world. The original project was funded by the World Bank, later by USAID. My assignment was to aid the Afghan Ministry of Higher Education (MoHE) on quality improvement and modernization, working directly with Deputy Minister Professor Mohammad Osman Babury (DM). The Deputy Minister oversaw the day-to-day operations of post-secondary education.

Professor Babury was someone I heard a great deal about from colleagues prior to arriving in Afghanistan. During my Kabul briefing when I arrived for this project, I was told even more about him and his work with the *Higher Education Project* (HEP) designed to improve the quality of university education. I was eager to meet this man who had eliminated rampant corruption in the university admissions system—a massive problem—requiring a $1500–$2000 bribe for admission to medicine or engineering, the two favorite choices, in a country with a $600 per capita annual income. Babury championed high-quality teaching and increased access to tertiary education, especially for women. The Deputy Minister was working with the Minister, Mohammad Azam Dadfur, to admit women once again to higher education. They had been removed from all levels of education except a few in the Medical School to treat wives and daughters of the Taliban. The Minister was also concerned with the high level of Post-Traumatic Stress Disorder (PTSD) among students—almost 80% at some point during their studies—the highest in the world!

The Higher Education Project (HEP) director, Michael Blundell, arranged for me to meet with the Deputy Minister at the Ministry of Higher Education on my third day in the country. Knowing that my success in the project depended on the Deputy Minister and developing a good working relationship with him, I became so nervous about the meeting I felt pains in my chest as we approached the Ministry.

We drove to the Ministry (MoHE) from our fortified and purposefully inconspicuous residence and office in Kabul in an armored vehicle, accompanied by my driver, my assistant, and an armed guard required for all American project members by USAID. We passed through the near total civil war destruction of one-third of the city from the front of the Ministry for at least ten miles to the wreckage of the former Queens Palace. The landscape appeared as if an atomic bomb was dropped. A few people were trying to live amidst the rubble—mostly in the remains of the ground floor or basements. Nothing stood higher than one story of pre-war multi-story homes and shops. Where a bit of the buildings remained, windows and doors were replaced by sheets of corrugated iron which those who survived scavenged, or by old rugs. In most cases all that survived was gray rubble.

Upon arrival at the MoHE, my colleagues and I were stunned by the good condition of the building in contrast to the devastation we witnessed on the way there, although hundreds of scars remained from bullet holes on the walls and a few large holes from more powerful shells. The compound was shaded by big, beautiful Eucalyptus trees around the edges and in the center courtyard, in stark contrast to the desolation in front of the Ministry. After checking our names, as well as the underside of our vehicle using mirrors on long polls looking for bombs, the guards opened the steel gates to let us into a large parking lot surrounded by a tall stone wall. We parked in the nearly empty lot that was now, for security reasons, used by known VIPs or people with appointments.

I later learned that the Ministry building and the University of Kabul next door were spared attacks during the civil war because the warring factions negotiated there regularly and wanted the facilities to remain in one piece for their safety. Eleven "armies" surrounded Kabul at one point—half of them allied with various foreign governments including Pakistan, Iran, China, Saudi Arabia, Russia, and later the US. The other armies were linked to particular political leaders or warlords.

The inside of the Ministry building was austere—grey walls, in various states of disrepair, a garden and rusty fountain in the center courtyard, which had not been cared for in years. Lots of people walked briskly through the hallways, most in traditional Afghan dress—long pants and oversize shirts for men, and several layers of long skirts for the few women among them. A uniformed guard showed us to the entrance of the Office of the Honorable Deputy Minister S. M. Babury.

My assistant Mujtaba and I entered the DM's outer offices. The two rooms were packed with at least 25 people patiently waiting to see him. The waiting room felt cold and reeked faintly of sweat and smoke from their clothes—though no one was smoking inside. I was immediately recognized by the DM's assistant whom I met before on an earlier visit for a different World Bank project. He immediately ushered me into the DM's office alone. The office was recently painted white and still smelled of fresh paint. Otherwise, the walls were bare except for the obligatory picture of President Hamid Karzai.

DM Babury immediately rose from behind his desk, excused his assistant, and shook my hand with the usual male two busses on the cheeks in greeting. I felt his sharp whiskers—a two-day growth is regarded as proper. The Deputy Minister spoke near fluent English and was working on improving his level of proficiency. I waited to greet him since official protocol required that senior officials at this level always spoke first.

"I have been eager to meet you, Professor Hayward, ever since reviewing your vita a few weeks ago and talking to others about your previous work in Afghanistan. I know a lot about you," the DM laughed. "Welcome to Kabul—once the beautiful Middle Eastern and Asian wedding capital of the world—a little past its prime today."

The Deputy Minister was tall, thin, about 50 years old, balding, distinguished-looking dressed in a dark blue Western suit and matching tie. As I learned right away, he treated everyone kindly while listening to their stories. And if there was a decision made, he took great care to explain the policy and the reasons for his determinations. He was much like a priest or teacher with a crowd waiting to consult him each day. Yet by the time he left, usually at 7:00 p.m., each of the petitioners was heard and most left satisfied even if they did not get what they wanted.

The Deputy Minister was a strong family man in spite of his long hours. He would often send his driver to pick up his six-year-old son at school and bring him to the office where he would "put him to work" drawing pictures. His sub-teen age daughters would often come too and work on their homework while he held meetings or spoke with petitioners. Both girls spoke perfect English often correcting their dad. Since I was soon regularly invited to dinner, and often proceeded to the DM home, the girls would come into the living room with tea and chat with me. They were quite articulate—ten and eleven going on 21. Professor Babury was Sunni, and his wife was Shia (not unusual in Afghanistan)—they were third cousins as is the custom in Afghanistan—picked for each other by their parents. The DM and his wife clearly developed a strong and powerful love for each other.

The girls were eager to read more books in English so my wife, a linguist, picked out age-appropriate books in English which I brought for the girls on each trip. I usually worked in Afghanistan for eight to ten weeks and then went to the US for two plus the month of August. The girls read them voraciously. Although men and women are normally kept separate, the girls were allowed to talk and read with me, until they reached puberty. Over time they would greet their "Uncle Prof." with a hug. His wife, a medical doctor, would greet me (which was unusual for Afghans—as visitors usually did not see the wife) and then retire to the rest of the house including the kitchen to cook for us.

Madam Babury was an excellent cook and always fixed multi-course meals discovering my favorites and fixing them often. However, if I were invited along with an Afghan, she would not come out and greet us. If my driver and guard stayed while we ate, Madam Babury would send food out for them. Although we never talked about it, I knew his wife did not greet us on those occasions because Prof. Babury needed to preserve normal Afghan customs to maintain "respect" given his high position. We ate sitting on cushions on the floor. It would only be the men (including his six-year-old son), with the daughters serving and bringing with them the wonderful, varied smells of food from the kitchen. We would eat from a huge multi-colored tablecloth spread out on the living room floor with many plates of food. The house had no chairs as was typical of Afghanistan.

As our conversation continued on that first day. The DM said, "You are exactly who we need here Professor Hayward, with your experience in accreditation, quality improvement, gender equity, and strategic planning. I want to start with accreditation. Can you start today?".

"Yes," I responded, totally flattered by the instant rapport that seemed to have formed between us.

"Good," he said, "To be fair, let me start by telling you a bit about myself. I was born in Herat near the Iranian border. When I was fifteen my father (fɑ ðər), who ran the regional telephone system for the government, was arrested by the Taliban as a potential subversive. My mother and I were sure this was a mistake since my father was totally non-political, though he maintained views—which he never expressed. When the Taliban police left with my dad, I followed them on my bike at a safe distance thinking the arrest was all a terrible misunderstanding and they would let him go when they arrived at the police station. Instead, he remained in jail for almost a year. Without my father around, or his salary, my family suffered during this period; although neighbors and relatives helped us a great deal."

"During the Russian occupation of Afghanistan, I excelled in school and earned a scholarship to an elite Russian university finishing first in my class, mostly of Russians. The party leadership at the university tried to force me to join the Communist Party; even though I refused they allowed me to finish my studies.

Returning to Kabul after graduation, I taught Pharmacy at Kabul University. Soon after arriving, I protested Taliban presence and interference in the operation of the University. That resulted in my arrest and time in jail. I was eventually released, as with my father, with no trial. Thereafter, the Taliban watched me almost constantly."

Listening to the Deputy Minister was captivating. I felt as though I was not in the office of the Deputy Minister of Higher Education in Kabul, rather more a close friend chatting while riding a bicycle with him. Although it was winter, the office, unlike the waiting rooms, was fairly warm because of a wood space heater well stocked with eucalyptus wood. A faint smell of wood smoke was in the air—pleasant, with a bowl of lavender herbs simmering on top of the heater which was vented into the wall.

"You have gone through many terrible experiences. Sir, I'm amazed they didn't derail you along the way," I commented.

"Most Afghans suffer from some degree of PTSD (Post Traumatic Stress Disorder). During the civil war, I lived in Kabul with the warring factions shelling from the hills surrounding the city. That is why you see one third of the city destroyed in front of the Ministry and beyond. We set up armed neighborhood watches because soldiers from one of these armies would come into the city to kidnap girls for sex slaves and young men to fight. A number of good ex-soldiers from the war against the Russians lived nearby, and everyone possessed guns (mostly AK47s) in our area, and thus the troops of the warring factions were usually afraid to enter. Nonetheless, many of my neighbors were killed or wounded when they ventured out for water or to find food. My brother was wounded one night by a stray bullet when fetching water for us. We spent most of our time in the basement and witnessed a good many of our neighbors murdered. This was a horrible existence: little food, no working toilets, not enough water to bathe, constant fear of shelling; yet no place was safe and escaping from Kabul at this time was almost impossible."

"Dr. Fred, 'Jan' [dear friend in Dari[2]], many people are waiting in the outer office for me, and I want to talk to a few of them while you read the policy document prepared for the World Bank. The paper is written in English. I would appreciate your reaction," he emphasized.

Then Professor Babury buzzed one of his two assistants, one man, one woman, and asked the man to send in together the first five people in line. The group included one policeman, a couple of university students, and an older man with just one arm bringing his son who appeared about eighteen. The Deputy Minister stood and greeted each of them warmly and listened to their individual requests: the first for help with a MoHE boss who made him work twelve hours a day; the two students came with a request for transfers to other universities; and the young man who came with his father, seeking admission to medical school. The DM asked his other assistant to call the MoHE police chief to his office immediately. Professor Babury was among the few in the MoHE who hired women as part of his efforts to improve opportunities for women.

The first student sought a transfer because he was planning to marry in Herat and wanted to attend that university with his wife and be near their family. His colleague desired to accompany his friend. The DM wrote, "OK to transfer to Herat University" on a piece of paper, signed it, and handed the paper to the student getting married. He said flatly, "No" to the other.

Then he asked the father who was wearing a traditional dress—a very long traditional woven shirt and loose pants, and looking distraught, "What brings you here?".

The father replied furiously, "My son has not been admitted to the university and I demand your help getting him into medical school!" He stood up waving his good arm. The policeman, in his crisp blue uniform,

[2] Dari is closely related to Persian and is one of the major and two official languages of Afghanistan along with Pashtu. Other smaller language groups are Uzbeki, Turkman, Hindi, and several others.

who accompanied the group, was so worried about the aggressiveness of the man that he jumped up and took several steps closer in case he needed to protect the Deputy Minister. However, the man quickly sat back down.

The DM asked the student, "What was your admissions test score?".

The student said, "About 200."

"That is way below the minimum for admission to medical school which is usually at least a score of 310. The minimum for admission to any program in a university is at least 220," the DM calmly explained.

The Deputy Minister went on to say that admission was based on two factors: student choice of institution and score on the entrance examination, with subjects such as medicine having higher score requirements than others such as education.

The father angrily fumed, "I fought the Russians and lost an arm and suffered other internal injuries. I have not been able to work because of my wounds. The country owes me a great deal beyond the medal they gave me. I think on the basis of my sacrifices; you should admit my son to medical school!".

The DM rose and heartily embraced the man, thanking him warmly for his bravery, "We owe you a great debt for your service." He then called in another assistant who clearly knew what the DM wanted from long experience, since she entered prepared with a stack of papers.

The DM asked, "Where did this young man rank on the entrance examination sheets?" The assistant found the name on about the fourth page and showed the Deputy Minister, "You are about two hundred people below the last applicant admitted to medical school by score." He turned to the father, "What do we tell those two hundred students who would be bypassed if we admit your son this way?".

The father stumbled a bit. "They would object and complain that the decision was not fair."

"Yes," said the DM, "I say to your son, you have two chances at the admission examination. You used one. Go and study hard and achieve a high score next time to honor your father and be admitted on your merit and make him proud of you." Father and son left the room with smiles on their faces—appearing satisfied.

The conversations were quietly translated to me from Dari to English by my assistant. At the same time, I was trying to read the document the DM gave me—with limited success. At this point the MoHE Police Chief arrived, saluted, and stood at attention.

The Deputy Minister asked, annoyed, "Why are you making this officer work 12 hours a day when 8 is the legal limit?".

The chief sputtered trying to explain, "We do not have enough officers to cover security around the clock, so we require our men to work longer hours to protect the Ministry."

"Why didn't you come and tell me that? We could fix things" the DM queried, puzzled?

"I was afraid to," the chief admitted, apologetically.

"Never be afraid to tell the truth about a serious problem such as this. How many staff do you need?"

Still standing stiffly at attention, the chief said, "Five more police."

"Here is a note to the human resources office to allow you the five additional police."

Then everyone else departed. The DM and I returned to work. My head was full of thoughts about the various requests. I was amazed at the amount of time the DM devoted to every person realizing they were potential presentations to the whole community about the actions he would and would not take. Separately they would tell many others about the visit. The time spent with each would get the message out about the visit almost word for word. President Karzai had told Babury privately that he felt he needed to write a letter of support for a low-scoring student, to send with a potential student who was the child of a VIP, knowing that the DM would turn the request down. "Keep it up!" the President had advised, "That way I make a constituent happy, although I know my recommended action might not occur." Karzai's conversation gave the DM courage to continue to be tough. On the other hand, I wondered how he managed to get through all the supplicants waiting in his outside office each day—which he did.

"My goal is for you to help facilitate a higher education strategic plan for the whole country for the next five years," the DM began. "However, first we must develop a strategy to solve the critical issue of the lack of women students. The Taliban banned all women from education at every level, kicking out most of them from the university, leaving about 380 women students and half a dozen female professors in the medical school over the years they were in power. They did not want men dealing with their wives and daughters. We also have a few Vice Chancellors who oppose admitting women. How should we approach this? During the last Vice Chancellor's meeting several VCs from the South argued this was a political issue which we educators should not be discussing. They believed this was a decision for the politicians to make. The majority disagreed, although discrimination against women in higher education is indeed a well-established problem."

At this point, four Vice Chancellors (VCs) and two Senior MoHE Directors the DM had summoned before I arrived, were ushered in. In addition, the MoHE Imam, my assistant, and interpreter joined us, along with more green tea and biscuits. All were seated on couches around the office with the DM in his usual blue armchair.

After the obligatory greetings and busses on the checks, one of the MoHE directors spoke first, "Islam requires that men and women should be separated. We must do that!".

The DM interjected, "That is not correct! As we emphasized in the MoHE's gender statement, Islam condemns any type of discrimination such as by age, gender, nationality, race, color, religion, disability, political views, or social status. According to Islam, men and women have equal rights in society. Islam calls for men and women to maintain relationships of respect, honor, and goodness."

A dissenting VC from the South, staring at me, interjected, "This is a foreign plot by the Americans to change us. We must not let them undermine our culture."

"No, that is not the case," said the MoHE Imam, "the Deputy Minister is correct, as directed in the Holy Koran, Islam states that men and women are equals. Many of us have worked toward gender equity for years."

"Remember," the VC from Kabul University said, "at one point, more women than men attended Kabul University. We can do that again. Hiring women faculty members will be harder because of the degree requirements and the small number of women with Master's or PhDs. However, we can start with the pool of those who used to work for us before the Taliban edict." Other heads nodded in accord.

The DM interjected, "Let us not get into an endless debate on this question. Gender equity is a settled issue as far as the MoHE is concerned. I would like us to hear from our good friend and colleague, Professor Hayward, who has a great deal of experience with gender issues in many other parts of the world including Pakistan and South Africa. Several of you met him two years ago when he was here working with universities. What is your view, dear friend?".

"Thank you, Sir," I replied. "The Government of Afghanistan signed an agreement recently at the gender conference in China to work toward gender equality, and President Karzai is committed to the agreement having promulgated a Presidential Decree to that effect. I think you should take the problem head on in a meeting with Deans, Vice Chancellors, and senior staff emphasizing that gender equity is now Government policy—everyone must work toward gender equity. When I was here before, many parents told us they definitely wanted their daughters to get a higher educated yet would not send them to the university without safe, proper housing."

"Do you think a few of the embassies would help us build women's dorms?" the DM inquired of me.

"I don't know. When we were exploring the possibility of private universities for the World Bank, we kept hearing from donors and embassies that 'they no longer do bricks and mortar.' On the other hand, I have already received criticism from two donors, the UN Women's Office and the Canadians, that the MoHE was not building dorms for women. I know you don't have money for essential construction such as dorms or laboratories, both of which are desperately needed."

"No, regrettably we do not have such funds!"

"Given those problems, I could write up a short paper stressing the need for women to have access to higher education, as they did before, to foster national development, and to stress the need for safe, proper, supervised dorms with the hope that donors will change their minds and fund new dormitories for women."

"Would you do that?" the DM asked, clearly pleased with my suggestion. "I think coming from a well-known foreigner such as you would strengthen the case."

"Yes, I will write that paper right away," I said, "I will lay out the case and will do my best to embarrass the critics by noting, 'You complain about the lack of women in higher education. Without your financial assistance new women's dorms can't be built. If donors don't help build women's dorms, they cannot complain about the failure to admit more women into higher education.'"

"That is an excellent idea, Dr. Hayward."

DM Babury also wanted to start work at the same time in setting up accreditation. He said he had invited several senior faculty members to his office the next day to talk about it and asked if I could give a short presentation then about accreditation which he knew I had worked on for the World Bank in a number of other countries including Pakistan and South Africa.

The next day he and I met with about half a dozen senior faculty members, most in tradition dress, sporting long beards, but clearly pleased at being asked to the Ministry to discuss an important issue. I gave a short fifteen-minute presentation about accreditation suggesting that it was an important way to improve quality by clearly setting standards for higher education institutions and then using peer review after an appropriate period of time to see if institutions met those standards and if they did grant them accreditation.

The initial reaction was mixed with some saying this is not the Afghan way. One suggested that: "Afghans are not willing to speak critically about any one or any institution. Peer review will not work in Afghanistan. Another though Afghanistan did not need such a "foreign import" since the current quality was excellent. Others thought it was a good idea but were worried about the cost and time involved. In the end they agreed that they would be willing to look at an outline of what I thought should be done and help the DM then make a decision about whether or not to have accreditation.

I did that and we met the next week. Interestingly, the fact that both India and Pakistan had established accreditation agencies, and their thoughts that this would be a way to obtain improved quality in those "other" universities that were not as up-to-speed as they were. Thus, with the ascent of the Minister, we embarked on setting up the accreditation process and in the end included both public and private institutions in the effort. It proved to be very popular with many institutions as a way to improve some of their weaker departments. In addition, although several private universities were strongly opposed, as one said: "let the buyer beware," other private institutions saw this as potentially an important recruitment tool if they were to be accredited and asked to be evaluated as soon as standards had been set. The process was quickly established, and peer review began—that being a much slower process.

June 26, 2003

The Plight of Women in Afghanistan

To understand the experiences and aspirations of Afghans, we carried out focus groups with secondary school males and females (11th and 12th grade), teachers, college students, business people, professionals, women's groups, and others. We started in Kapisa in the far north of Afghanistan to talk to secondary students and teachers. Later Razia Karim, who worked with the Deputy Minister and me, did a survey of girls at Kabul University which laid out the kinds of problems they faced both at home to get permission to go to the university, and at the university with sexual harassment and other problems.[3] Her survey emphasized that in spite of MoHE efforts, many problems still existed. She is another excellent example of the people who made things happen.

Even with the changes in laws about the treatment of women, we were struck by the extent of their continued inferior status. The high school girls wore burkas when they came to school and stored them in their backpacks while on campus. They put the black garments back on, including covering their faces, before they left. Asked by my colleague Sara Amiryar, why they wore the burkas, they said, "Because truckloads of boys will beat us with sticks if we do not." Sara had always refused to wear a burka.

Sara Amiryar, a US citizen of Afghan origin with an MBA from a US University, was my partner in Afghanistan on my first trip. She worked in the admissions department of a Washington DC University, was active in Afghan Affairs, and had been elected to the Loya Jirga[4] in Afghanistan a few years earlier, which laid out the groundwork for the new constitution and civil rights including for women. She was fearless, outspoken, and bright—as well as a joy as a colleague because of her knowledge, good humor, and optimism. Sara had recently worked in Afghanistan on a project for secondary schools and had excellent connections in Afghanistan from both that project and because of her family history in the area which allowed us tremendous access, access which proved essential to meet our goals in the six weeks we had on this trip.

Sara was treated with great respect and often deference even in places where she must go off with the women. At times she was allowed to join the men—the only women, because of the work she did with children two years earlier and because her father was a famous and loved minister in a previous government and was singled out and killed by the Russians when they invaded Afghanistan. People told us that Afghanistan had much greater gender equality at least in the cities (such as Kabul and even more conservative Herat) before the Taliban. In those days

[3] See: Hayward, Fred and Razia Karim. (2019), "The Struggle for Higher Education Gender Equity Policy in Afghanistan," *Education Policy Analysis Archives*, vol. 26, no. 10.

[4] A traditional kind of Afghan Parliament called together to discuss major issues—in this case a new constitution for Post-Taliban Afghanistan.

the Constitution was enforced with a guarantee of equality for men and women.

Our team included one other Afghan staff member with an MBA from Pakistan. He was wonderful at putting our data into tables and charts and excellent in dealing with computer problems. A driver had also been assigned to us.

We scheduled focus groups in Kapisa. The trip from Kabul to Kapisa took hours on a terrible road badly damaged with bomb craters. The hulks of dozens of burned-out Russian tanks and armored cars littered the road along the way. This was a major escape route for the Russian retreat at the beginning of the 1990s when the Taliban (with Ben Laden) and war lords defeated them—continuing to ambush the troops as they departed for Russia and inflicting heavy casualties. In spite of the road conditions the trip was fascinating—with beautiful countryside, small towns, and acres and acres of farms, now green with crops.

The discrimination against women even in the capital city Kabul when I arrived was the worst I have ever seen. In many of the non-urban areas that were extremely socially conservative, the discrimination was universal, often in the "name of" Islam. The blatant discrimination was ubiquitous and was especially galling for Sara who is a strong proponent of equality for women and a major force in Afghanistan for women's rights. She had also been an effective spokesperson for Afghanistan in the US and been interviewed by McNeil Lehrer, CBS, and others.

Sara rightly emphasized that gender discrimination is not what the Koran says, in spite of that frequently given justification, and quotes several passages from the Koran emphasizing the equality of men and women. She did not hesitate to make her opinion clear, while respectful of traditions. For many people women do not count as a whole person—at best half—as the Vice President of Afghanistan told us, having recently looked at the pilot of the national census and election registrations. He found in several regions the ratio of men to women was as much as one woman to four men even though a much larger number of men than women were killed during 25 years of war.

When the Vice President queried the officials in charge of the pilot census in these areas, he found that just one of the census takers had been surprised enough to question this ratio of men to women. When he did question the results in several cases, the male heads of the households would say women and girls do not count as one person—at best one half if at all. Thus, the undercounting of women was tremendous in that area with the ironic effect that Government funds and most aid funds were reduced since they were based on reported population. Undercounting women by a factor of four reduces that area's funding enormously. Since this was a pilot study preceding the actual census, the Vice President insisted that there be recounts in those areas with follow-up questions whenever the gender ratio seem skewed. The VP was a former judge noted for his fairness and honesty.

The blatant inferior status accorded women impacted our work tremendously making much of what we needed to know difficult to obtain. On several occasions, Sara was greeted and expected to join the other women in a different house both for meals and talk. She warned me in advance this would happen, and she argued (correctly, I think) that this is not the place to debate the issues. She insisted, however, on equal status as a member of the Loya Jirga and gave an interview with the BBC, at that time—which made several people extremely angry.

Gender equity is decreed in the Interim Constitution, although those pronouncements are not enforced in practice. As one Kabul woman lawyer told us, you can put equality in the law, however, if that is not part of tradition, or not enforced, the law will not work. When I asked if one could sue businesses, universities, or individual men in the nation's courts for sexual discrimination as we do in the US, she reported that if any women—including herself—ever tried to sue a man, agency, or company over gender discrimination they would be "laughed out of court" by the judge. She said, "no judge anywhere in the country would hear of it." This lawyer was bright, articulate, and attractive (which she noted can create problems). She was a senior prosecutor at the Ministry of Justice!

The projects Sara led for the Academy for Educational Development (AED) the previous year included the "Bluepack" project by AED and the White House—US school children contributed money for 160,000 blue backpacks for secondary students. The backpacks included pads, pencils, school supplies such as a small blackboard, chalk, a ruler, etc., along with a thermos—all with appropriate "Gift of USA" and the AED logo on them. We saw a good many of these Bluepacks on children as we traveled outside Kabul, and Sara was greeted at any school we visited with great affection and respect. Working for AED helped us a great deal under these circumstances.

In Kapisa we carried out several focus groups. This was the one place we needed to have separate focus groups for men and women—and even there it was the school teachers, not the 11th–12th graders or the parents who wanted that. Unless we talked with the teachers separately, there would have been no such interviews. Even in liberal areas such as Kabul, secondary school boys and girls are taught at separate schools, or at least separate compounds with walls in between.

In spite of the sexism and discrimination against women, in Kapisa Sara was generally treated as an equal. She was especially effective in trying to right abuses of women when we came across them. Two days earlier she learned about the kidnapping of two young girls (13–14) by one of the Northern War Lords (one of the War Lords of the sort the US has supported). The motivation for the kidnapping was probably to force a marriage or perhaps as revenge for something done by someone in her clan. We were in the car at the time Sara received a cell phone call telling her about the kidnapping. She immediately phoned the Minister of Interior whom she knew. He had recently heard about the kidnapping and sent police (they are national police) to try to rescue the girls. That would not be easy since these war lords operated independently of any law with their own armies. Many were selling heroin—as they were when the US supported them with guns, missiles, mines, money, and advisers against the Russian—in spite of knowing they were in the drug business. This time the girls were rescued.

Sara found herself a target because of her strong actions and comments. A group of Mujahideens tried to kill her for speaking to the press during the Loya Jurga by ramming her car with a truck loaded with rocks. The crash came close to killing her. She still limps from the injuries to her legs and back. Her driver was killed. Add to that all the conservative men, the sexism, the fact that many people have guns (usually long rifles or Kalashnikovs) made the situation often dangerous for Sara. We understood she was constantly at risk. However, as she said, if someone does not stand up for the rights of women, who will and how will the problems ever be acknowledged or rectified? A few men are working for gender equity (including President Karzai, the Minister and Deputy Minister of Higher Education, and Minister of Interior). Nonetheless, this is an uphill and difficult battle, especially now that the Taliban are back in power and reversing most women's rights.

Housing for Women

At the request of the Deputy Minister, I wrote a five-page piece pushing for proper, secure housing for women students. It was widely circulated by the MoHE to all the donors and embassies which had provided funding in the past. I made the case for encouraging women to attend higher education and for the necessity of good, secure, supervised housing for women who did not live in the university town. I pointed out this was a key to national development and that the country could ill-afford to waste the talents of half the population.

To my amazement a few days later, I received a call from the Cultural Affairs Officer, Jean Manes, at the US Embassy. She had arrived two days earlier from the US. She asked, "Could you come for dinner with me at the embassy this evening? I read your document about the gender issues in Afghanistan, and I want to talk to you right away. I have made all the security arrangements for you to pass through the barriers, and I will meet you at the entrance at 6:00 pm."

I quickly organized my own travel and security arrangements (the obligatory armed guard) and received permission for our crew, my driver and guard, to enter the embassy compound. On arrival at the gate, several plain clothed US security people were waiting for us. They hustled our vehicle to park in a secure area and led me into the embassy with none of the usual body checks, no giving up my cell phone—the VIP treatment! The Cultural Affairs Officer was there waiting.

Jean Manes was an attractive and well-dressed professional with short nicely styled hair. I later learned that she and her husband, Hector Cerpa, were both excellent cooks when they invited me for dinner the following week. Jean was earnest, clearly hardworking, and focused on getting things done.

Jean started off saying, "I am finally getting to meet you. I was afraid you might have trouble with our security, so I sent officers out with your picture, and I see you make it through all right. I want to start by asking how you became interested in the gender issues?".

I responded, "This has been a concern of mine since I started teaching at the University of Wisconsin and helped start the women's studies program, defended women's equity in hiring, and reviewed women's salaries finding huge differences which were clearly fostered by gender. As a dean, I was given the power to address the problem and did so.

Becoming aware of the extent of the problems, I have worked on gender issues ever since."

We walked to one of the Embassy dining rooms where Jean's husband waited. The smells emanating from the buffets were rich and surprisingly enticing. I noticed they were serving lots of different types of ice cream for dessert including several of my favorite—rocky road, chocolate fudge, and more. I was hungry. We chose our dishes from the vast array of tempting options. Jean never stopped asking me questions—talking from the moment I entered the gate. She could not believe the US Embassy had done nothing for women in Afghanistan recently.

"Even worse," I said, "a year ago I brought up the terrible problems women faced getting into the university, obtaining funding, finding safe proper housing and, once they attended the institutions, dealing with harassment. I urged USAID to support gender equity efforts. I was told the Embassy was not interested in the gender issue—that it was 'too sensitive' for the US to become involved. I was dumbfounded."

"Well, that is going to change!" she stated, emphatically.

I later learned that Jean Manes was a close friend of Hilary Clinton (then Secretary of State). When Secretary Clinton visited Afghanistan, she was especially impressed by the young university women she met. Secretary Clinton promised to help women become empowered including assistance with campus housing. The appointment of Jean as the new Public Affairs Officer was part of her effort.

Jean and I enjoyed a delightful conversation. She hardly touched her food and took copious notes. She asked, "Where do you think we should build the women's dormitories? I am thinking of three facilities."

"Politically you must build two in the provinces, one in Herat and the other in the North. You should also build one more in Kabul since the city has a disproportionately high number of students from outside the area—especially women students sent to Kabul from the countryside to allow them to gain a quality secondary education so they can pass the entrance examination that will allow them to attend a university. The US built a large dorm here years ago. The new dorm could be on vacant land next to the existing dorm so that facilities can be shared and should be open to all non-Kabul resident women students at any of the four public higher education institutions and universities in Kabul. We will badly need the new space given our effort to increase the number of women students from 375 when I arrived in 2009 to about 10,000 (18%) in three years. In the new strategic plan, the goal was 50% women by 2020.[5]"

"OK, let's meet again in three days, which will give me time to examine the situation here. We will go from there, Fred. Now I must leave for a meeting. I am stubborn, so I guarantee we will get these dorms built."

Jean Manes was indeed a doer! Her husband walked me to the exit while we carried out a brief conversation—after I finished my rocky road sundae.

My five-page memo worked wonders. NGO donors, the French, Indian, German, and several other embassies, along with the narcotics agency, agreed to build eight excellent dorms for women across the country. Work on the first US dorm would start in three months, since the US Embassy already had plans from the last women's dorm construction, the Army Corps of Engineers was available, and the land was available. This began an important new era for higher education. Jean Manes was at the head of this new effort organizing not only the US Embassy and US military, but the Germans, Indians, and several others. I met with her almost weekly during this period.

She did the unheard of for US Embassy personnel in Afghanistan. One evening she invited Deputy Minister Babury and me to dinner with her husband at their apartment plus another senior official. Again, all the guards asked if I was Prof. Hayward, and when I said "yes" they ushered us in without further questions even though the DM forgot his ID since his security guard had put it in his briefcase in his car and we left in mine. When we realized that he was without the paperwork, I called to warn Jean of the possible problem on the way, and she cleared the hurdles. Professor Babury and I were given the VIP treatment and zipped through security. This was especially important since the previous time the DM visited the US Embassy he was pulled aside, and body searched. He vowed he would never return—no US diplomat was treated like that when visiting the MoHE or even President Karzai. I told Jean that story and talked the DM into giving a visit to the US Embassy one more try. It went ahead without any problems.

[5] That turned out to be unrealistic given the many problems. The actual figure in 2020 was 27.8 (MOHE records), nonetheless a great improvement given the still limited housing for women and the opposition of the Taliban and other groups to university education for women.

The five of us spent a memorable evening of discussion, laughter, and wonderful food fixed by Jean and her husband. We also carried out critical policy discussions about what the Embassy could do to help women at the universities as well as support the MoHE's efforts to enhance the quality of higher education in general. Here, the Deputy Minister made a strong impression on Jean with his knowledge and thoughtfulness. I believe the evening marked a turning point in the relationship between the MoHE and the US Embassy.

A downside to my work on gender (housing for women, the Gender Strategy, and the Sexual Harassment Policy) was attention from the Taliban. While I never gave press interviews and stayed out of the limelight, I was present at the televised meetings of the Vice Chancellors once a month with the Minister and other senior staff. Being the sole non-Afghan there, the TV cameras focused on me frequently to see my reactions to various statements by the Minister or Deputy Minister zooming in on my face, even when talking in English. The Taliban were still destroying girls' schools, though they had not attacked the universities at this point.[6] However, they knew that I was working on gender issues with the MoHE, the women's group of MPs, and faculty members.

A few days before I was scheduled to go on leave, the DM planned a farewell party for me starting with an award at the Ministry followed by dinner with special guests at his home. The awards ceremony was quite moving with speeches by the Minister and Deputy Minister. At the ceremony, a long Afghan robe was presented to me which I put on to great applause and laughter. More than a dozen women in the audience came up to thank me at the end of the program—quite unusual for Afghanistan.

The DM and I then retired to his office for lunch and to talk over a few last-minute issues before I left on leave. When we arrived at his office, he excused his staff and shut the door—which was unusual.

"I have cancelled the dinner tonight. Intelligence has determined there is a plan to kidnap you at this event. While the Ministry of Interior is prepared to send 25 well-armed troops to protect us, people will be killed. I don't want to run that risk, so I have cancelled the event. I am sorry."

[6] That changed in 2016 when the American University was attacked, probably by the Taliban, killing thirteen people including students, one faculty member, and several police. There was an additional attach on 2 November 2020 when 32 people were killed and more than 50 injured, also attributed to the Taliban.

I agreed with his decision—nothing is worth the death of innocent people. We finished our work and said our goodbyes. He sent me to my lodgings with an extra security detail. The next day, I made it to the airport and flew home to Washington DC. My assistant had not slept for the last two nights since he was sure I would be killed. "The Taliban always get their man," he said. Two weeks later, an American and a Canadian Professor at American University Kabul, a few blocks from our own office, were kidnapped in broad daylight right outside the university. Two years later, they were still held hostage by the Taliban who were demanding an exchange of prisoners and money for their release. I could easily have been among them. Only in late 2019 were they released in exchange for three Taliban leaders who were imprisoned.

On Being a Target

I had been a target before. I think about it once in a while when working in places that are war zones or where there are individuals and groups hostile to you because of your nationality, your color, your gender, your religion, or your perceived wealth. The problem first occurred to me when I was in India on a program designed to "build bridges of understanding" between Americans and Indians. Groups of people, mostly Communists, were actively anti-American at that time. And you realized when their demonstrations occasionally turned violent, that if you were there you could be hurt or killed. It would not be you people were after, rather what you symbolized to them. You could not gain absolution because you affected good intentions, because you cared, because you were a nice person.

You knew the hate was generalized, vague, unlikely to focus on you personally, yet the concern was always there, nonetheless, in the back of your mind. Twice I have been targeted specifically for who I was. That was while working with the South African Electoral Commission—and that was quite real. Lots of people were assassinated during this period in South Africa; the Electoral Commission building was surrounded by white reserve soldiers (at that time there were no non-white soldiers in the South African military) from Cape Town called up to guard us. Though they may have harbored doubts about the idea of free elections, they were professionals and carefully protected us. The other took place in Afghanistan.

Early in the preparation for counting in South Africa, I was surprised when told I was on a "hit list" because I was one of two people leading the counting process for the whole country. I was not afraid; I did not run through a list of things undone. The idea of an attack targeting me or Pirishaw Kamay, my partner in this, seemed foolish since we could be replaced. Apparently not in someone else's mind who saw virtue in our demise. And someone did manage to get into the Electoral Commission Office building with a hand grenade but was caught.

Those experiences do make one cautious. Yesterday I found a bottle of Jim Beam in a store in Kabul—much to my surprise given the laws against liquor except for foreigners. I bought the bourbon, pleased with my discovery. When I arrived home the bag felt damp and I discovered the seal was broken and that the bottle had been opened, that there was still far too much liquid for a factory bottled drink, that tampering occurred. Perhaps this was nothing more than a watered-down half bottle—half water, a scam. Yet it also occurred to me that perhaps I was that nameless target. Only Americans drink bourbon in Afghanistan. What a clever way that would be.... For a moment I thought about drinking it anyway, a passing idea was all. That idea did not make me any more or less afraid. Yet it did make me think about the mindless, anonymous nature of hate—irrational yet real, meaningful yet meaningless. At times I think I understand the origins of hate—in poverty, abuse, violence, torture, victimization, revenge. And then at other times I realize I do not understand it at all. I poured the liquid down the drain.

Little Things of Beauty

On a sunny day in 2003 in Kabul, I finished my work for the day and was walking back from the office, a multi-storied residence a few blocks from the guest house where I was temporarily staying. The other large residences also had fenced-in yards, such as our current guest house, in what were once the homes of wealthier residents, no doubt upper or middle-class families.

The war was still especially real here, and we heard jet fighters almost every night and occasional bombing on the outskirts of the city. Taliban rocket attacks on Kabul occurred from time to time. They did not have the range to reach our part of the city. People told me about the smart bombs now used by the US, and how they could hit a target and by going deep, minimize the damage to surrounding buildings. I did not believe that. I was told that the residence a few doors down from us was the former home of Taliban leadership and thus became the target of such a bomb. I was curious, and since I was in no hurry, thought I would take a look at the bomb site up close on my way home—a few hundred yards further down the road.

There it was an enormous hole probably two stories deep. Indeed, it was clear that the walls of the house had caved inward, and everything had fallen into the deep basement, yet there was little damage other than to the house. A few of the garden walls were cracked and one or two fell down, although the neighboring houses and their outside walls were still intact, and no windows broken. It was amazing!

Standing by the edge of this carnage, to my delight was the sight of green poking through the rubble. Amidst about two feet of debris, a rose was peeking out. I stood transfixed for a few moments. The rose bush displayed four or five branches covered with beautiful clean green leaves washed by recent rain. Elevated above the rubble stood a single flower—a single beautiful red rose—one the likes of which one saw in Afghanistan or Iran (the homes of the rose)—a perfect red rose, partly opened—the sort one could purchase in a florist shop in the US. Here, enmeshed in the carnage of a bombed-out Taliban residence in Kabul in the middle of a war—grew a rose that signified to me at that moment, the hope for rebirth in Afghanistan—the hope for a future of beauty, love, and success. It represented more than just survival, but the possibility of building something new and better even in the faces of war.

June 8, 2003

Survival

Walking to the guest house last evening, I witnessed a boy about 12 or 13 looking through the pile of garbage at the end of the street, turning over every bit of rotten, smelly, spoiled food, scraps of paper, and other trash. He was not even using a stick; rather going through the piles bare handed. He must have found a gem or two because he picked an object out of that pile and put it in his already half-full plastic bag.

I skirted the garbage heap each day because of the foul smell and hordes of flies. And over the days walking past, I was surprised to see that the city seemed to pick trash up each day, unusual in my experience elsewhere. Thus, the smell was not as bad as most such dumps. This nuisance and irritant for me was a major find for an Afghan child—sustenance, nutrition, and survival.

June 10, 2003

The Garden

Eating my breakfast in the garden of the guest house this morning, I was struck by the beauty of the flowers, the spectacular roses blooming, the well-tended grape arbor with young grapes maturing to be picked in another two months. Small beds of plants had been set out in several places in the middle of the lawn, carefully avoided by the customers while they grew to maturity and bloomed. The results of the care devoted to this bit of peace and quiet by the gardener seemed incongruous amid the hubbub of the city and the clamor of war on the other side of the wall.

The quiet and beauty were particularly striking because of yesterday's warning that intelligence agencies learned that three suicide bombers were driving around Kabul planning to bomb (the authorities knew not what). They were expected to focus on a site that would mar the celebration of the anniversary of the Loya Jerga (a kind of constituent assembly) on June 11th. Taking this all in, I noticed one of the five Turks staying here while working on development projects, leave a lively conversation at their breakfast table for a moment, get up and go over to the edge of the lawn where he spotted a tiny growth of vine trying to make its way toward the sky. He squatted down, while continuing his conversation, gently untangled the vine from the grass, and carefully wrapped all dozen inches of the vine around a pole holding up the arbor. Natural, touching, irresistible by another human being who shared the spell of this beauty.

June 11, 2003

Love Amidst War

Walking back from the office to the guest house for lunch, about 12:30 in the afternoon, feeling the sun hot on my back, happy for the little whiffs of breeze blowing in my face, I was trying to make my way along a well-swept sidewalk with periodic holes which I needed to avoid, between the deep drainage ditch and the road. I was also looking carefully for the garbage, and rubble strewn in front of a building destroyed by bombing and sand piled in anticipation of new construction. At this hour, the street itself was nearly empty—not alive with people as at other times. This was lunch time for most, nap time for many Afghans. A couple of school students were walking home in their blue uniforms, a worker or two, and a number of omnipresent guards were sitting in front of each building. They do not get official breaks, protecting offices and homes around the clock. Several of the guards lounged with their Kalashnikovs at their sides, squatting together in that typically Afghan pose that would kill my legs, with bottoms resting on the ground. A few guards sat in the shade of wooden guard houses in front of the high walls that surround every building.

Approaching the door of one of the guard houses as I walked, wide open, as was the window across from it, I heard the unusual sound of a woman giggling and the soothing happy words of a man wafting from a cot in plain view through the open door. Barely a few steps later, and a few feet away, did I realized they were making love, legs akimbo for the one, the other in tight embrace entwined on top. They were moving rhythmically for all to see, yet fully clothed in multi-layers common to both men and women even in the heat, yet oh so connected in ways concealed. Passing quickly and as invisibly as I could, I was struck by their joy, the happy sounds that were emerging, and the oblivion to the circumstances around them—guns, bombs, people, the public eye, extreme conservatism. I continued on my way surprised to see that I too was sharing a sense of their pleasure—not sensuous or even sexual—simply the joy of happiness and love.

Kabul and War

This World Bank project in Afghanistan was designed to help six universities develop strategic plans and proposals for a potential $5 million each—a small beginning to the actual sums necessary given the fact that 80% of the buildings were damaged, most of the sciences had no working labs, and only one of the universities had a computer lab with Internet connections. I am putting on three days of workshops in Kabul for the six university presidents and two staff involved in planning at the universities chosen to participate in the project. We are holding the meetings in the Ministry of Higher Education in Kabul. It gets only about three hours of electricity. Most of the bullet holes and broken windows have been repaired and a generator kicks in for the most essential functions when power is off. The rest of Kabul gets about three hours a day of electricity on a rotating basis—if they are lucky. My assistant says they get electricity about every other day for a few hours where he is staying.

The university presidents are excellent—in two cases in particular, a fundamental and encouraging change from my last visit. The presidents are working hard to understand strategic and financial planning and have been effusive in their appreciation of our effort. They will not receive Bank funding unless they put together excellent proposals for repairing and upgrading the universities. Their first efforts are, on the whole, excellent and most will get the initial $500,000 of the grant based on their proposals and my approval of them.

All of this progress was occurring within the context of war, a fact I knew; yet the last few days brought the war starkly to the forefront—a sad reminder of the hurdles ordinary people face and the courage it takes to lead in a higher education institution. A suicide bombing took place in Kabul yesterday and intelligence reports suggested that there was another car loaded with explosives roaming around looking for the "right" target and opportunity to blow it up. That Monday was the opening of Parliament—the first in about 30 years, and the Taliban had threatened anyone involved. Every Afghan trying to make this country work as a democracy and a developing economy, was a target—and of course—so are those of us who advise them.

Because of the bombing yesterday, we are required to remain in our compounds today (they are guarded by gun toting Afghan Ministry of Interior troops who have an excellent reputation). Six to eight of them protected me the last time I was here—but solely when we were outside Kabul. They put their lives on the line for us, and I trust them totally—especially after several harrowing experiences in which they put their bodies in front of mine.

In the face of the threats, this was a quiet and strange day in Kabul—I had seen nothing similar before. I went to the office to get my ticket for my break (with special dispensation since we are not to go outside) and I witnessed quite a scene. Almost no cars were on the road. Security was everywhere with machine guns at the ready, fingers on the trigger, walking in the streets, on buildings, in the compounds. This is the stark reality of war. All around were the armored cars, the troop vehicles, the helicopter gunships, the fighter jets leaving cloud-like trails high in the sky. The Taliban claimed responsibility for the car bombing yesterday, and intelligence expects more bombings through Monday when Parliament meets.

Parliament was to be sworn in on Monday. All senior Afghan officials would be attending plus the US Vice President Dick Cheney and other international dignitaries.

The Taliban have threatened to attack those who participate. All the Ministries are closed, not usual for a Saturday since Friday is the Sabbath and a day off. All the NGO workers were told to stay home, and clearly not many people were going outdoors.

I was not afraid, for I have long ago felt that when my time comes it will come—probably in an auto accident. I am careful, yet not frightened. We saw that the threats were not simply press stories, this was real. I walked the two blocks to the office with two armed guards with me and one outside the office.

June 20, 2003

Rural Afghanistan: A Day with an Extended Family

Friday is the "day of rest" in Afghanistan—a time when one goes to the mosque, visits family, catches up with friends. I chose to go to the office knowing it would be empty, to catch up on the tasks I dislike doing such as my "expense report" and "time sheet," and other unexciting chores essential to survival. No sooner was I settled in than my colleagues Sara, Essa, and Kareem came in asking if I would join them at an extended family gathering in a village outside Kabul. They did not have to twist my arm, and we set off at about 11:30 a.m.

The countryside along the way was exceptional, filled with lush farmlands, much of it irrigated, with some crops almost ready to harvest, others in their early stages. The mountains in the distance were stark and with the tallest still covered with snow. Along the road, we passed an incredible mix of people walking, cars, bicycles, goats, hand carts piled high with bags of cement, a few horses, a string of camels, painted giant trucks loaded to overflowing with goods, SUVs, buses, and more people. As we drove through the valley, we saw many small villages, the tents of nomads camped along the river, bombed-out houses and farms, as well as other structures under reconstruction. I did not realize how much bombing had occurred in the countryside.

I was struck by the incredible beauty of the countryside in stark contrast to the devastation we left behind in Kabul. Spread out in front of us was the flat landscape of the valley, the river winding through it with people washing vehicles, themselves, and their clothes. I was taken by the colors, the fields with their various hues of green leading to the hills which abruptly rose in steep rocky slopes—a few dark and forbidding, others soft and inviting. It was a wonderful contrast to the blast walls and machine toting guards of Kabul.

The Family Home

We turned off the main road and headed up a track to a cluster of houses perhaps 500 yards away, past fields of onions with their seed pods waving in the wind, lettuce, wheat—all in small well-tended irrigated plots edged by small embankments high enough to keep the water in, wide enough to walk on. The closest house was recently rebuilt given the destruction of the war in the area, and we could see huge windows open in an upstairs parlor that seemed to call out to invited guests. Around the large house were about half a dozen other smaller ones—several of which were not yet repaired—all made of that hard-baked brown clay that merges earth and house, yet seems impervious to rain. Odds and ends of small farm equipment were scattered around the yard: an old pump, a few rusted barrels, and an old Russian tour bus that was once someone's pride and joy. The bus was now propped up unceremoniously on bricks, one tire missing, the others flat, with the once elegant curtains at the windows still hinting at the glory of years gone by.

The men were ushered upstairs to the sitting room—the women went off to another house in the compound. Indeed, we did not see any of the women again until after lunch, and then only Sara. The sitting room was quite large, probably fifty by thirty feet. The floor was covered with multiple layers of Persian carpets and cushions propped snuggly against all four walls. The walls were bare and freshly painted, and a wonderful cool breeze blew across the room from the open windows on two sides. We took off our shoes when we entered, greeted, and shook hands with those assembled, starting with the patriarch. Five or six other men were seated on the floor as well as one or two young boys. We were later joined by several other men.

The patriarch was sitting in the middle of the longest wall in front of a window. He was turbaned, with traditional Afghan trousers and a long loose shirt. He sported a full grey beard, kindly eyes, and a welcoming greeting though he spoke no English. Sara's husband, Quadir, was my "guide" and interpreter. We paid our respects in ways not unlike other places. "How are things in Kabul?" "When did you arrived in

Afghanistan?" and "What was the work we were doing?" The other guests were all members of the extended family. One was visiting the family and his childhood home for the first time in 23 years. He had been living in Las Vegas for most of that period, working in entertainment, and was now a US citizen. He brought his wife and three children too—though we never saw them. Another lived in Kabul and ran a pharmacy, one was a driver, and yet another attended law school. Eventually about a dozen men were in the room including sons, cousins, a son-in-law, and several other relatives.

The family was gracious and seemed happy to see me. They had heard about this American working in Kabul. The greetings were warm and courteous—for a few even deferential with a two-handed shake followed by placing the right hand on the heart. A huge tablecloth was brought in, made of canvas, and placed on the floor covering the whole area between the cushions. The food was served—wonderful smelling dish after dish of food: yogurt, eggplant cooked in olive oil, plates piled high with rice, a wonderful dish of large meatballs in a delicious sauce, platters with tomatoes, onions, and cucumbers. The host provided glasses of milk, bottles of water, and a huge bottle of a "diet" cola (neither Coke nor Pepsi), that clearly was purchased for me based on previous intelligence about my diabetes. Plates loaded with hot freshly baked pita bread; several plates of freshly cut watermelon; a salad of tomatoes, cucumbers, mint, and other veggies kept coming. This was truly a feast to be relished, with more food than all of us could possibly eat. We sat cross-legged on the cushions around the tablecloth. Plates of food were passed and kept coming around throughout the meal until all of us were full. While we ate, we carried out a long series of conversations about Afghanistan and world affairs, translated patiently for me by Quadir, assisted at times by the brother from Las Vegas.

Conversations

Much of the discussion was about family, the beauty of the place, the crops about to be picked in the groves, the devastating effect of the five years of drought (especially on their apples) which was finally broken this year, and the usual banter and joking of a relaxed afternoon. People also had serious issues they wanted to discuss, and once they found out that I was neither reticent nor would I be insulted by tough questions, we began a long conversation about US foreign policy, President Bush, the war in Iraq, and the changing perceptions of the US in much of the world. The family and guests made it clear that they were grateful to the US for ending the horrors of the Taliban and for working to bring security to the area. They were feeling that it was finally possible to begin to rebuild their lives and to think about the rehabilitation of the country.

They asked about the recent US invasion of Iraq, and I told them why I was opposed. Their real question was: "Why would the US go against the judgment and wishes of almost all of the rest of the world when that invasion seemed unnecessary—why did the US put itself in a position where it was hated by much of the world?".

I told them, "I do not have a good answer for that. You know what President Bush thought. I considered the action was a mistake and might well have been done collectively in a relatively short time with international support and little or no violence."

The patriarch, who did most of the talking, recalled how the US had been loved widely in earlier years and built a dam in Afghanistan that benefited many people. He emphasized, and others agreed that: "Now the US interests seemed self-centered, focused on oil and business." The patriarch continued, "What kind of a person is George Bush?".

We then carried on a fascinating discussion that demonstrated a broad knowledge and understanding of much of the rest of the world. All this from someone who spent most of his life either farming, albeit a medium sized once-prosperous farm, or fighting. Indeed, I was struck by the fact that this extended family seemed to know more about American politics than a similar random sample in the US.

We turned to the issue of the war in Afghanistan and the destruction that could be seen almost everywhere. People talked about the years of war, the depression and heartache caused, the need to flee to the mountains on several occasions, later to Pakistan, and for some of them, asylum in various other parts of the world. The patriarch himself fought against the Russians. He emphasized, "Our frontal assaults failed against the Russian tanks and guns; Then we learned to use guerilla tactics with great success." He also noted the problems created by, "factional fighting among the various Mujahideens groups, the corruption and venality of many of the warlords, and the destruction they caused in their greed, especially in fighting over the 'spoils' in Kabul when the Russians were defeated."

The patriarch and the others expressed great gratitude to the US for liberating the country from the Taliban and the terrorists—for "rescuing us." They said they were honored to have me in their home—that in fact I was the first American to visit their home. People also worried that the US was not going to stay long enough to ensure stability, that we would abandon them once again. The patriarch also said that the rise of the Taliban should be a warning to the US, that the injustices, inequalities, repression, victimization, economic deprivation, and lack of education rampant in the world, breeds and feeds these movements. He hoped that the US would return to an era in which it had the rest of the world at heart, would once again see people as equals, and once again become committed to development for everyone.

I was struck by the wisdom and the thoughtfulness of those gathered in that room—especially that of the patriarch and several other elder statesmen of the family. I was reminded again of how often I have been in villages and towns in Africa, talking to what my Ghanaian students used to refer to as "our poor unfortunate illiterate brethren," yet when I would go to the rural areas, I found them better informed than my students. Here too was more wisdom than I found in many a classroom.

After the dishes were cleared off by the younger men, we were ushered outside for tea in chairs set out on a huge carpet in the shade next to an irrigation trench flowing with water. After a while we were asked if we wanted to take a walk up the hill to see the orchards—and the younger of us did. The trees grew on terraced plots started decades ago by the great grandfather of our host. This was a large farm—at least 100 acres, I guess. Parts of the orchard grew apple trees, plus cherries, mulberries, peaches, and apricots. In a flatter part of the upper terrace, potatoes grew with their shoots now up a good eight to ten inches. We did not venture beyond the edge of the grove for many landmines were still planted there from Russian airdrops—some identified by red flags, others waiting for the unlucky.

We enjoyed the view and the cool breeze. Conversation stopped for a while as we took in the quiet, bird sounds, and the voices of distant travelers on a path along the hill. Soon one of the sons and several of the children came up the terraces carrying a huge cloth. One of them climbed a nearby mulberry tree, the cloth was laid out, and to peals of laughter from the children, it was soon raining mulberries. A basin of water was produced, berries and hands were washed, chairs mysteriously appeared, and we sat for a while enjoying the fruit, the orchard, and the warmth of wonderful company.

This was an afternoon of unrivaled hospitality, a wonderful chance to be part of an extended family gathering—to talk, to exchange ideas, and to learn. The day was a great privilege for me, opening a chapter of life in Afghanistan I never thought I would see.

People write that the Afghans are a closed book. I do not believe for a minute that is the case. Perhaps too few people take the time to listen and share ideas.

August 16, 2010

Checking Up on the Kankor (Admissions) Exam with the Deputy Minister

To move our agenda forward faster and to deal with the seemingly 100 odd projects we have embarked on together, I spend every day with the Deputy Minister as a sort of temporary (and foreign) unofficial chief of staff. The work is interesting and gives even more access than I had before. The two of us have also become friends in the process, which makes the long hours easier. We often work until after 7:00 p.m. in the evening. The Ministry of Higher Education is one of the few Ministries that has been able to accomplish much recently, largely due to Minister Dadfur and Deputy Minister Babury. Both of them are amazing, deep thinkers, politically aware, while not being politically partisan, with strong values of honesty and hard work. The two are totally committed to higher education, willing to talk to anyone, and yet neither of them will suffer fools or crooks gladly. The Minister is particularly concerned about getting more women into the university and about student mental health. He is a psychiatrist. As a refugee in Pakistan, he established a successful mental health clinic in the major Afghan refugee camp there.

Today was a typical day for the Deputy Minister and me. We dealt with the breakdown of an agreement between Kabul University and an American University which had not lived up to its contract, USAID issues, nominations for five short-term fellowships to a World Bank conference in Thailand on governance and finance, appointment of a medical advisor, a leadership program, proper wording for a graduation certificate, some financial issues, and the Kankor (admissions) examinations which were being given today. This exam is administered by the Ministry of Higher Education (MoHE) and is the key for admission to all 23 universities and higher education institutions, plus private institutions, foreign scholarships, the military academy, and other higher education organizations.

Today was the examination for entrance to night school which operated at five universities nationwide.

I spent the morning with the Deputy Minister, the Chancellor of Kabul University, and several MoHE officials checking the Kankor night school exam process. We visited the examination centers in Kabul, followed for a while by a gaggle of reporters (at least 20 of them) trooping obtrusively through the exam rooms—acting like typical reporters, snapping pictures of the students taking the exams, talking to each other, shoving to get a better place for a photo, and even asking the students questions in the middle of the exams. Nonetheless, the faculty members and the exam takers were happy to see the Deputy Minister there since everyone knows he is the guardian of fairness and transparency. Indeed, in one examination room a student told him he thought a question was wrong in that none of the options for answers was correct. The DM pulled out a piece of paper and checked the question. Sure enough, the student was correct, and the DM made sure that question was not counted, and other test centers were notified of the error.

Six thousand people were taking the exam today for 1100 places, 5000 of them doing so in Kabul and the rest in Herat, Balkh, and Kandahar. We visited about 25 testing rooms at Kabul University, with the DM checking a few IDs and talking to students during the breaks about where they were from, if they found the exam hard, and encouraging them. In all the rooms we visited, students were seated in alternate rows with an empty space in between with at least 3 faculty members proctoring the examinations. For many of the students, this was their second chance—as they were allowed to take the entrance exam twice to improve their scores if so desired. I was awed by the organization and the security efforts. The whole campus was closed to everyone other than the faculty members, exam takers, and those of us involved in overseeing the exams.

After going through the first several exam rooms, the Deputy Minister held a press conference outside, which was typical for him, so as not to disturb the process. Questions were asked about the process, the number of women involved, and a number of issues about higher education and the strategic plan. The press conference was orderly and thoughtful with about 15 different networks involved. The press is free, alive, and very well informed in Afghanistan in spite of the fighting. They seemed pleased with the interaction and the DM's answers.

Then we visited a number of other examination rooms. At the end of the day, we toured the labs in the Pharmacy Faculty—the DM is a

Professor of Pharmacognosy and Phytotherapy in the Faculty of Pharmacy at Kabul University, so this was his "home." The labs were impressive, a few with modern equipment from Japan and Germany and a lot of old equipment as well as drugs and medications going back to the 1930s. A few of these agents and other chemicals were still being used given the inability to afford new materials. The budget for the whole system of 23 public universities and higher education systems was $35 million. That would not even run the social science departments at the University of Wisconsin, Madison.

As we moved through the examination centers, the Deputy Minister told me his exam staff received word that about seven 2nd and 3rd year students were part of a ring, organized by two people working in the Ministry (no doubt soon to be unemployed) who were taking the tests for other people. They had photos and fake IDs. The organizers even chose individuals who resembled the people for whom they were taking the exams. The Deputy Minister gave the photos of those involved to the exam security and to all the proctors.

One of the students was caught early, another fled as we headed for the Agriculture Faculty. We were present later when another suspect was apprehended, and he was then questioned in front of us in a faculty lounge. He was a cool character. We did not believe his story. Nonetheless, I was interested to learn the final outcome since the interrogation continued after we left. Later two students were caught using cell phones to get answers to questions. The security people caught their accomplices, too, by tracing the phone numbers. The MoHE staff were impressive in their ability to keep the cheaters out. Yet most of the students were honest and interested in doing the best they could on the tests. During one break, the DM talked to the students about attempted corruption in the process, urging them to turn in anyone they found cheating. He noted that cheaters, if successful, would get their spots illegally and deprive others since the total number of places was fixed by capacity. The students agreed.

About 500 potential women students took the exam at Kabul University, which pleased the DM and the Chancellor of Kabul University. They noted that it was hard for women to get permission to be out at night and thus to take night classes. This made it especially hard for women who worked to get an education. The Chancellor had made a great effort to light up the campus at night with funds from a donor—the German

Embassy. He showed us an area where the NGO was installing lights and decent sidewalks.

We spent a busy several hours at KU. I saw more of the campus than I had ever seen before. They have expanded facilities with help from donors, although they still need much more space and enhanced building upkeep. They have a long way to go to meet the standards set for them at the MoHE. Kabul University has a beautiful spacious campus with lots of trees (which once was typical of all of Kabul) which were not cut down during 30 years of war since the place was occupied by the military which prevented that. In addition, unlike the area around the university, the campus was not shelled because this was the agreed meeting and negotiation place between the various sides in the civil war and they all agreed to protect the buildings. Numerous meetings were held there to try to end the conflict or at least to agree on rules of conduct for the fighting. The campus is not similar to the "real world" of Kabul although it is an indication of the park-like atmosphere that once prevailed throughout the whole city. The campus offers a cool, shady haven from the now real world of Kabul which has blast walls in front of every building. Inside, beyond the walls, in most cases, lie beautiful gardens, which illustrate why Kabul was once a place people came from all over the area to be married. The gardens are still extensive with lots of roses blooming most of the year and other flowering plants everywhere.

December 2013

Forces for Change

Many people other than those mentioned above put their lives on the line to bring about major positive change. Professor Farid Danish of the Medical School, who agreed to lead the review panel on private higher education institutional quality was another of these. When he agreed to lead the study, he told me, "This may get me killed but I will do it because it is the only way to ensure quality higher education in the long run."[7] He did receive threats and he and his team had to be accompanied by the police on some of the review team visits to private institutions. The review demonstrated that a number of the private institutions were below minimum standards set by the MoHE, as we expected.

[7] Personal conversation with Professor Farid Danish in early December 2013.

They often had political connections or were run for-profit with little interest in quality higher education. Eighty-three private higher education institutions were reviewed, with thirty being below minimum standards—thirteen of them ranked very poor with recommendations that they be closed.[8] The then Minister, involved in the original approval to open some of them, refused to close more than eight of the units rated "very poor" and quashed the report of the review. Many other academics and employees, including several Vice Chancellors, Ministry Staff, students, and faculty members, who worked hard to improve conditions in higher education, were threatened and even assaulted.

The amazing transformation of higher education in Afghanistan, with the addition of accreditation, broad improvement of program quality, upgrading of teacher training, protection and addition of women, and upgrading of facilities, succeeded in part because of outstanding leadership, but particularly as a result of the broad commitment of the university community to change and improvement and support of the community. This collection of people, at every level, made it possible to create the transformations that took place from 2002, at the end of the Taliban Period, to the present time even under the most difficult of circumstances. One critical result was to move from nearly no women in higher education to over twenty percent. That has since been reversed with the return of the Taliban to power.

That broad understanding and agreement on the changes that were needed, at every level, in the university community, plus the willingness of people like Deputy Minister Babury, Professor Farid Danish and Frozan Abide to put their lives on the line, made transformation in higher education possible, including for women. The question now is whether or not these positive changes will be sustained with the withdrawal of US and NATO forces. Already, the Taliban and other terrorist groups are targeting people who worked with the allies as interpreters or employees and those educated in the West, on a daily basis. The allied governments have realized the dangers to these people and promised that they would be given visas to the relevant countries, Great Britain, Germany, and the US, but that process has been terribly slow sometimes taking three years to process an application. That does not leave enough time to prevent the slaughter of some of those who helped the allies by the Taliban. And of

[8] Hayward, Fred. (2016), "Report on the Review of Private Higher Education Institution," MoHE, Unpublished report of July 16, 2016.

course, it does nothing to protect the ordinary citizens from the wrath of the Taliban and other terrorist groups. Equally importantly, this departure may lead once again to create a situation in which US or allied forces will need to return, though at the moment there is little stomach for that. We have kept US troops in South Korea since the 1950s. I think it would have been equally prudent, and cost-effective in the long run, to have kept a small force in Afghanistan as a deterrent. I am not optimistic about the future without that. What we have seen to date is keeping girls and young women out of most education, despite Taliban promises to the contrary. The country is suffering from a drought, loss of funds from donors until women's right and other conditions are met, and the seeming lack of concern of the Taliban about the starvation of millions of their people, especially children.

The post-Taliban return is in its early stages and thus how it unfolds is yet to be fully clear. However, in spite of promises about an enlightened Taliban today, they have failed to date to allow girls education beyond primary education, freedom of the press, and women to work (within their definition of Sharia Law), and thus their start to date is not encouraging. Reporters have been beaten up (and then a pro forma apology given), women's rights demonstrations violently broken up by their troops, girls kept out of schools, while boys were told to go back in mid-September 2021, men and women are to be taught in separate classes at universities—which the Taliban suggests could be done by every other day classes for women and men, though that is not in place at the time of writing. In addition, in late September the Taliban enunciated what the New York Times called "'Spine Chilling' New Rules for Media in Afghanistan," which effectively shut down more than 100 local media and radio programs and caused many journalists to flee for their lives or go into hiding.[9] In addition members of the Taliban have killed some police and former soldiers in spite of promises not to do so.

No women have been involved in their new governing body, and travel continues to be restricted. While a few flights have been allowed to leave with Afghans for Qatar and a few other places, I suspect that is largely to encourage European and American countries to release blocked funding. To date, in addition to the defective organization of the withdrawal of US and NATO forces, mostly caused by the lack and/or incompetence

[9] Gall, Carlotta. (September 24, 2021), "'Spine-Chilling' New Rules for Media in Afghanistan," *New York Times*, p. A11.

of US planning, to date there have been few efforts by the allies to live up to their promises of continued humanitarian support for the millions starving in Afghanistan or to help get the remaining stranded Afghans, and a few US and European citizens still wanting to leave, an opportunity to leave. In the meantime, the economy of Afghanistan is collapsing, and the country is facing more drought and massive starvation.

The overall effects are a moral and psychological disaster for the US that will not soon be forgotten. While the United Nations and World Bank have initiated some excellent food and humanitarian programs, much more needs to be done. In the meantime, both the Biden and European institutions have turned to other matters. This is a tragic moment in US, European, and Afghan history for which many of us are deeply ashamed.

CHAPTER 7

Conclusions: Reflections on More Than Fifty Years of Working in Africa and Asia

In the six cases examined above—Sierra Leone, Ghana, Madagascar, South Africa, Pakistan, and Afghanistan—I have tried to show, through examples, how successful transformation, especially in higher education, is not only the result of excellent leaders, but based on critical individual public participation at all levels—people who work to create the conditions for success. These efforts in the successful cases, helped established an environment which allowed for change, especially those things that facilitated the establishment and support of democracies allowing for debate, decision-making broadly, and the resolution of differences. Where that did not happen, as in Sierra Leone after military intervention, major changes thereafter were usually difficult if not impossible, although there is some recent hope after new elections. Too often we leave out the role of individual citizens when we look at attempts at, or successes in, development, and focus primarily on leaders. I have tried to highlight both leaders and ordinary people who have placed critical roles in the change process. These cases demonstrate the importance of transformation of individual actions at many levels. In what follows, I highlight the most important lessons from each of the six cases.

F. M. Hayward, *Fostering Institutional Development and Vital Change in Africa and Asia*,
https://doi.org/10.1007/978-3-031-04364-2_7

Ghana: Collaboration and Success

Ghana is one of the few African countries never to have a civil war. This has helped produce a strong sense of community locally and nationally as well as great pride among Ghanaians. At the same time, a traditional commitment to local and regional autonomy and responsibility remains and that too has been important to success. That is clearly expressed by the achievements in the two local cases examined in the chapter on Ghana: that of the University for Development Studies (UDS) in the North, and in Apiride in Southern Ghana with its traditions of healing and cooperation.

The UDS was underfunded by the government from its inception in 1992, yet in a relatively short time was built into an outstanding example of student, and community-centered high-quality education led by the Vice Chancellor and Deputy Vice Chancellor, faculty members, students, and especially support from communities in the North. The university flourished because of the home-grown assistance from regional groups, chiefs, local governments, and individuals. Local governments gave the UDS unused building around the Northern Region (mostly failed 2-year technical academies), land, and money to facilitate growth. Because of the student-centered program, UDS attracted funding from the Carnegie Corporation and later other donors. Its curriculum, as emphasized in the chapter on Ghana, was a unique innovative program focusing on local development using student groups during their summer vacation—called the "third semester program." Development projects included clinics, schools, bridges, and new types of drought resistant agriculture, among others, which were developed and implemented by students and local citizens. This credit-bearing program, with a paper in the final year, created an amazing rapport between the UDS, local communities, and student teams working jointly on the projects. The projects were developed collectively with the students, faculty advisors, and local people. The students and locals advanced the projects together over three years, each group raising the funds needed for the project, and working with the local communities during three "third semesters" to implement them.

I interviewed students, local participants, and faculty members about this program, and they all reported accomplishment of the major goals of the projects—a clinic completed, a primary school built, a bridge constructed. Sometimes success required major revisions in plans to deal with unforeseen problems—all of which were worked out between the students and local communities, creating a new understanding of local conditions, higher education needs, and local cultures. Building UDS was a remarkable community effort in support of regional projects, helping the university upgrade its programs, expand and prosper, as well as lobby for more government support. Indeed, its long-run successes attracted more government and donor funding.

Similarly, the accomplishments of Apiride, in Southern Ghana, in solving local conflicts that had prevented traditional healing, was another example of traditions that fostered an effective community environment locally and collectively in society writ large. The importance of civic agreement and collaboration was illustrated by efforts to deal with infertility in Apiride. Once a resolution of the conflict was achieved by the new chief, harmony returned to the community and allowed traditional medicine to function effectively. The power of traditional medicine in Apiride was recognized throughout the region especially for patients with severe broken bones that hospitals could not fix, sending the patients to the one and only specialist in Apiride where that family had generations of achievements with broken bones. In many respects, that specialist's training as a child was similar to what takes place in a contemporary medical school including preparing medicines, in this case based on herbs gathered in the forest which were effective in killing pain and improving healing. He also was taught a deep understanding of the human bone and muscle structure. Comparable successes took place in Apiride with mental health and other problems using local specialist, traditional methods, and local techniques.

Striking was the amount of "science" in most of these efforts and the critical nature of belief in the process for both achievements in mental health and infertility. These publicly available amenities helped create a strong sense of community, well-being, and comfort in Apiride. Multiplied throughout the country, these successes and sense of community produced the kind of national environment that fostered democracy, national development, and contributed to education improvement and success. They are an excellent example of how local leadership and public participation foster development and change as well as their critical importance to success.

Sierra Leone: Democracy Versus Corruption and Military Rule

Sierra Leone was the first place I visited in Africa in 1966 when working on my dissertation. At that time, Sierra Leone was a budding democracy—one of the first in Sub-Saharan Africa to have an opposition party win an election. When I arrived, I noticed the openness, the excitement of people about the higher education system, the high expectations for development, and growth of the economy—expectations which were being achieved at that time. Sierra Leone had an excellent education system at every level—especially in the urban areas. The two higher education universities were among the best in sub-Saharan Africa, attracting foreign students especially from Nigeria and the Gambia.

Shortly after I arrived, I met several extraordinary leaders in Sierra Leone, the most striking of whom was Siaka Stevens, head of the opposition All People's Congress (APC). Stevens was one of the most astute and wily politicians I was ever to meet, as noted earlier. Stevens, unknowingly, taught me how little I knew about African politics in spite of my years of undergraduate and graduate study of Africa in the US and England. Stevens was open to my every question, allowed me to travel with him during the 1967 election campaign, which the APC won, and continue to meet with him periodically over the years gaining an intimate view of both the electoral process and the operation of the major political parties. Stevens attracted huge crowds wherever he went because of his political achievements and plain talk. He was a great storyteller with a wonderful sense of humor which he was able to share. The elections showed me the remarkable level of public understanding of national politics even in the most rural areas—indeed, better than the average citizen in the US because of the direct effect of government on their daily lives.[1] Stevens successes were also a result of his ability to attract local leaders (including chiefs), women leaders, and local citizens in large numbers.

I later came to know Joseph Momoh, President after Stevens and picked by Stevens. President Momoh asked me to come to Sierra Leone to help him and his team figure out how to return the country to a multiparty democracy and end the one-party state set up by the Albert Margai

[1] See: Hayward, Fred. (1976), "A Reassessment of Conventional Wisdom about the Informed Public," *American Political Science Review*, July 1976, pp. 433–451.

government. President Momoh succeeded in opening multi-party competition moving toward free elections when his efforts were undone by an invasion from Liberia and then a military coup.

Sadly, after an excellent political and economic beginning, the coup against President Momoh in Sierra Leone in 1992 led to a series of additional coups and crises within the military. Corruption became rampant, led by the military. The economy deteriorated, and despite the eventual return to democracy years later, that government failed to improve conditions or eliminate corruption. Education, in the meantime declined terribly and higher education became a shadow of its former high quality. Sierra Leone, unlike Ghana, Afghanistan, and South Africa, has not enjoyed the kind of public return to a commitment to democracy that might have turned around the political system. It demonstrates what happens when the public is not involved in change policy and the national sense of community that existed before the coup was lost during military misrule. During that period, the public became cynical of government—an opinion that has not changed in recent years. Sierra Leone is now at the bottom of many indices including the highest under-5 infant mortality rate, a literacy level of 43%, and low-quality education.[2] Though the economy has improved slightly in the last few years thanks to mining, the democratic ethos has not been rekindled though the last national elections may bring about some positive change. Unlike South Africa and Afghanistan, at the time of writing there were no groups actively pushing for reform, improved education, or effective democratic government. All this is a sad commentary on what was such an outstanding start in the immediate post-colonial British era. In part it demonstrates the long-term havoc resulting from military rule, the toll of widespread corruption introduced by the military but continued by the elected governments, in part a lack of community commitment to democracy, quality government, and other institutions.

South Africa: Unification Versus Attempted Fraud

I was asked to return to South Africa, where I was working with twelve historically underdeveloped (Black) universities, in April 2004 to help with the national elections. The African National Congress (ANC) had

[2] UNICEF. (2020), "At a Glance: Sierra Leone Statistics," downloaded on May 7, 2020, from: https://www.univef.org/intobycountry/sierralene_statistics.html.

gained reliable intelligence suggesting that the governing National Party, and the Inkatha Freedom Party, were planning to try to fix the results of the first multi-ethnic national elections in history. Knowing that I had worked on, and written about elections in Africa, I was asked by Frank Chikane (who I had met when he visited the US), head of the South African Council of Churches and a member of the Election Commission, to join them to identify the problems and help prevent this potential fraud.

My partner in this effort during the elections was Pirishaw Kamay, a major figure in the trade union movement, and one of the few in the majority population to have election experience through union elections over the years. Indeed, ANC fears were correct. We discovered that the system established to count ballots would not work, a lack of needed equipment for counting, three million missing ballots on the first day of voting of a three-day election cycle, sub-routines to change results in the standalone Election Commission Computer, and efforts to submit fraudulent results from fake polling centers during the counting. In the end, with the help of lots of ordinary citizens, and hardworking students, we succeeded in ensuring fair elections by identifying these problems and eliminating them.

This transition period showed the tremendous potential power of a repressed majority population to force a government willing to use extreme violence, imprisonment, murder, and restrictions on movement for the black population, to eventually give up power. Agreement on goals among this diverse population during the struggle, aided by a sizeable number of white supporters, fostered the changes.

Many people worked for years to end apartheid. Among them was Professor Sibusiso Bengu. When I first met him, he was Vice Chancellor of the University of Fort Hare—one of the "historically Black" institutions I was working with on strategic planning. His commitment to equality and his lack of fear in the face of threats, was amazing. While I worked with him, two attempts were made to kill him by Inkatha thugs, one of which nearly succeeded. On another occasion, Vice Chancellor Bengu stood in front of the Ciskei Police at the entrance to the university to prevent the police from coming on campus to harass students and no doubt injure or kill some of them. When he became Minister of Education, he asked me to serve as an advisor. His openness and willingness to listen was part of his success.

For me, one of the most amazing experiences during my time in South Africa was to see the democratic processes at work in the ANC, whether with youth groups or at higher levels. One of the ANC rules was that at a meeting everyone who wanted to speak received a chance to do so; no one could speak a second time until all who wanted to had spoken once. On several occasions, leaders interrupted, but were always told to wait their turn until all others who wished to had spoken—and they did! I had never witnessed this level of participation anywhere else, yet I saw this process repeatedly in South Africa. Consequently, some of the meetings were prolonged—but no one was left unheard who wished to be and that included supporters of the Apartheid Government—a remarkable example of people's democracy.

Madagascar: Agreement Not Decrees; Equality Versus French Arrogance

My work for the World Bank in Madagascar was, until the end, one of the most enjoyable and satisfying for me up to that time. Much of that was due to the exceptional leadership of Haja Nirina Razafinjatovo, Minister of Education and Scientific Research and his support by the President of Madagascar, Marc Ravalomanana, as well as the community at large. The eventual willingness of the Vice Chancellors and faculty members to embrace reforms once they realized how far behind Africa and the rest of the world Madagascar was at that time, made a huge difference leading to success. Haja was effective due to his calm attitude, his determination to make the case for change by agreement not decrees, his careful thinking about change, and his ability to establish excellent relationships with people from primary and secondary students to heads of universities. Failing at first to make his case, he worked intensely all day, and often into the evenings, to figure out a more effective strategy to get his points across. He would come in some mornings with a diagram and outline of a new approach and try the ideas out on me. We would talk about this new approach, sometimes make changes, and then work some more. In the long run he succeeded in vastly improving primary and higher education. Haja also assembled an outstanding staff of female and male educators, especially at the primary and tertiary levels, to help develop and implement plans. He also loved jazz and traditional Afghan music.

A surprising aspect of Madagascar was the high level of equality of women—especially among the educated. This tradition grew out of a long history of powerful women queens but was also reflected in a general attitude that at least educated women were equal to men. This did not hold true in some rural areas. Nonetheless, even in such places, discrimination against women was nowhere nearly as devastating as in rural Afghanistan or Sierra Leone. Several women were in senior positions in the Ministry including the head of Primary Education for the country, with a few in leadership positions elsewhere including Vice Chancellors. I have never worked in any other developing country where so many women were in recognized positions of authority.

On the other hand, in no country in which I have worked, except perhaps Senegal, has a foreign power and ex-colonial government, France, retained such power over its former colony or assumed that it had the right to influence, intervene, and sometimes control government. French interference was largely responsible for the coup that brought democracy to an end, stopped the improvement of education, and reversed some of the progress already made. These efforts forced President, Marc Ravalomanana, to flee to exile in South Africa. Haja and other Ministers went into hiding when the military took over. The coup led to a reintroduction of corruption on a massive scale by the military—which President Ravalomanana had succeeded in limiting before the coup. Madagascar had previously been among the most corrupt countries in the world prior to his presidency. The post-coup military took to confiscating people's cars, TVs, and other possessions, soon joined by the police and appointees to office. Even by 2020, the country had not recovered from the damage done to the economy after the coup, with a loss of tourism, investors, and trade. Despite extensive public support for President Ravalomanana, French involvement with the military and support of unrest against the President, as well as actions of the populist movement led by the former mayor of Antananarivo, brought progress and democracy to a crashing halt. This was a chilling example of what Robert Kagan has called "charismatic authoritarianism,"[3] which can lead, as it did in Madagascar, to disastrous political and economic results. The case also demonstrates how dangerous military intervention can be especially with foreign support, add to that uninformed populism and that can effectively ruin democracies, as it did in this case.

[3] See Robert Kagan. (Sept 26, 2021), in a superb piece on authoritarianism, "Our Constitutional Crisis Is Already Here," *New York Times*, A 28–30.

Pakistan: Top-Down Transformation of Higher Education

Pakistan presents a particularly interesting case in that, unlike its former co-dependent colonial partner India, almost nothing happened to improve education from independence in 1947–2002, in spite of broad recognition, and several studies, that demonstrated the poor quality of education and near bottom ranking in the region. In 2002, with strong support from President Musharraf, and outstanding leadership at the Higher Education Commission (HEC) working with the World Bank, it succeeded in making major transformations in higher education. These included sending several thousand faculty members for PhD training at home and abroad, upgrading the curriculum, building state-of-the-art regional science laboratories, and increasing access while keeping student-faculty ratios low and relatively constant. This was done self-consciously in a top-town manner with little consultation with higher education leaders or information to the public. It worked because of strong support, including major increases in funding, from the President and the World Bank as well as total commitment of leaders of the HEC. However, starting in 2008 and continuing through 2010, opposition to the HEC grew in government and some higher education institutions, as well as Parliament and several ministries, as noted above. Funding was cut substantially, and higher education once again was among the laggards in the region, with little academic or public protest. Pakistan is an excellent example of the risks of top-down change efforts and the consequences of transformation without fostering broad participation and understanding.

Afghanistan: Public Dedication and Support for Transformation During War

The achievements of the struggle for improved education, gender equity, greater access to education, and a high-quality higher education system during more than thirty years of war, is one of the most remarkably successful transformations of higher education I have witnessed. This progress was made while a war raged. War killed more than 100,000 civilians from 2009, when the UN began recording casualties, to 2019,[4] with at least 5000 Afghan troops killed in addition in 2019 alone according to President Ashraf Ghani in a radio broadcast. About 3500 US and other coalition troops were also killed in efforts to defeat the Taliban. In spite of ongoing bloody wars over more than 30 years, phenomenal progress was made in several areas including health care and higher education. This happened because of the dedication and support of large segments of the population, the bravery of many of its citizens, as well as the courageous leadership of a number of people at many levels, willing to put their lives on the line for positive change.

Over the eight years I worked with the Ministry of Higher Education in Afghanistan, I was impressed by the critical role played by some former students and faculty members, in addition to Minister Dadfur and Deputy Minister Babury—people such as Frozan Abide who graduated at the top of her master's class in Education and after teaching for a few years was named Deputy Vice Chancellor (DVC) of one of the universities near Kabul. She worked tirelessly to improve the level of instruction and insisted that faculty members upgrade their preparation and respond to student questions. DVC Frozan was also willing to fire those who failed to show up on time or at all—and there were a few of those—and thus made a number of enemies. She was threatened by local clerics who did not believe a woman should have such a high position, her office was burned, and she was menaced several times by opponents with AK47 machine guns.

The Ministry provided strong protection for her and so did most of the students. Nonetheless, this pressure took a terrible toll on her. I was assigned to be a mentor for her over two years and we met often to talk about strategies and next steps. I was always cautious about pushing her to continue when she felt like quitting, fearful lest she be killed. She never stopped, arguing that if she quit, the bad people would win.

Among the most important drivers of change in higher education in Afghanistan was Mohammed Osman Babury, Deputy Minister of Higher Education, at that time. He was a quiet, yet effective voice for

[4] *Al Jazeera*. (2020), "US War in Afghanistan: From 2001 Invasion to 2020 Agreement," p. 3 downloaded on April 24, 2020 from: https://www.almazeeera.com/indepth/interactive/2020/02/war-afghanistan-2001-invasion-2020-taliban-deal-200229142658305.html.

improved quality and gender equity for women. He was the most effective higher education leader I have worked with anywhere over fifty years for his thoughtful planning, his ability to gain support for higher education goals, for quality improvement including gender equity, and his success over ten years as Deputy Minister in bringing about major improvement to higher education. The achievements resulted from the fortuitous combination of Deputy Minister Babury's efforts plus those of individuals such as Frozan Abide, Minister Dadfur, Vice Chancellors, Deans, and countless students, faculty members, citizens, and donors, that allowed such transformation to take place from 2002 to 2016 during a bloody and devastating war. Especially important in Afghanistan was that higher education was isolated from politics in general, though with strong support from both Presidents from 2002 to 2016. As such, people were able to make changes in higher education that could not, at that time, be made in society as a whole, especially in terms of gender equity.

###

What is striking about the six cases explored in this book is the importance of broad agreement on goals at multiple levels, a willingness to run risks to achieve them as we have seen particularly in the cases of South Africa and Afghanistan, and a broad sense of community that recognizes challenges and fosters solutions to them as demonstrated particularly in the cases of Ghana, Afghanistan, and South Africa. Added to that is the importance of strong leadership at many levels, able to mobilize the public, deal with critics, raise needed funding, and organize the implementation of plans. All of the successes examined here involved extensive cooperation and teamwork throughout the change process including broad public support. We have also seen the critical role women have played in all the successes, both as leaders, and mobilizers, often willing to take major risks in the efforts to foster national and local development as well as gender equity.

These cases also illustrate the importance not only of planning but implementation. In all six of these cases there were strategic plans. However, in only four of them were they implemented. The failures were partly the result of military intervention in two of the cases, but also because of the lack of broad support from the public.

What I have tried to show here is the importance of individual citizens in creating the conditions for change, fostering agreement on goals, and taking part in implementation of their goals. It is the collective efforts of citizens, leaders at various levels, and the creation of a community locally that is responsive to the problems and the needs of its citizenry that leads to success. We saw this especially in the difficult environments of both South Africa and Afghanistan—cases in which tremendous transformation took place and led the way for broader changes in society including progress toward gender equity, increased access to education, and set the stage for other problem solving as new issues arose. While not all such efforts achieved success, especially in the cases in which there was military intervention, even in the cases that failed, the basis for future improvements often remained, to be revived at some point, hopefully in the not-too-distant future. We are yet to see what will happen in Afghanistan under the Taliban, in spite of its success in transformation during the previous twenty years. As of this writing the prospects do not look good. The successes, on the other hand, should give others hope for their own transformation in the future.

This study has been an attempt to go beyond the focus on leaders and the mere mention of "followers," without also focusing on the important contributions of "followers"—citizens at all levels—to programs that led to change. As we have seen, these individuals have often put their lives on the line for positive change, mobilized citizens, fostered the resolution of conflicts, and in many ways provided the ideas that led to successful change and development. As I said at the outset, the work of Amartya Sen (1999) which emphasizes the importance of the link between freedom and successful development, is borne out by much of what happened in the successful cases explored here. It is the importance of effective participatory processes which, as Sen notes, enhances the capacity of the public to be an active part of change and transformation, that needs to be emphasized, along with the critical role of outstanding leadership.

Bibliography

Al Jazeera. (2020), "US War in Afghanistan: From 2001 Invasion to 2020 Agreement," p. 3, downloaded from April 24, 2020, from: https://www.almazeeera.com/indepth/interactive/2020/02/war-afghanistan-2001-invasion-2020-taliban-deal-200229142658305.html.

Altbach, Philip. (2011), *Leadership for World-Class Universities: Challenges for Developing Countries*, New York, NY, Routledge.

Boervolk Radio. (2009), "I Love South Africa but I Hate My Government (2009)," downloaded on July 23, 2019, from: https://iluvsa.blogspot.com/2009/11/no-retribution-for-mbeki.html.

Burns, James Macgregor. (1978), *Leadership*, New York, Harper Perennial.

Gall, Carlotta. (2021), "'Spine-Chilling' New Rules for Media in Afghanistan," *New York Times*, p. A11.

Glickman, Harvey. (1992), *Political Leaders of Contemporary Africa South of the Sahara: A Biographical Dictionary*, Westport Connecticut, Greenwood Press.

Hayward, Fred M. (1976), "A Reassessment of Conventional Wisdom about the Informed Public," in *The American Political Science Review*, July 1976.

Hayward, Fred M. (1987), *Elections in Independent Africa*, Chicago, Illinois, Westview Press.

Hayward, Fred. (2017), "Lessons Learned from Strategic Planning for Improved Teaching and Learning in Developing Economies," *Planning for Higher Education Journal*, V45N4.

Hayward, Fred M. (2020), *Transforming Higher Education in Asia and Africa: Strategic Planning and Policy*, New York, SUNY Press.

F. M. Hayward, *Fostering Institutional Development and Vital Change in Africa and Asia*,
https://doi.org/10.1007/978-3-031-04364-2

Hayward, Fred M., and Daniel Ncayiyana. (2015), "Challenges of Graduate Education in Sub-Saharan Africa." *International Higher Education*, no. 79, Winter 2015.

Hayward, Fred, and Razia Karim. (2019), "The Struggle for Higher Education Gender Equity Policy in Afghanistan," *Education Policy Analysis Archives*, vol. 26, no. 10.

Hilal, Hilalluddin. (June 12, 2003), Deputy Minister for Security and Police in a Meeting with the Author and Sara Amiryar.

Kaburise, J.B.K. (2003), "New Variations on the African University: The UDS Experience with Developing an Alternative Approach to Tertiary Education." Reprinted in the *African Journal of Education and Practice*, 2017.

Kagan, Robert. (September 26, 2021), In a Superb Piece on Authoritarianism, "Our Constitutional Crisis Is Already Here," *New York Times*, A 28–30.

Kahn, S. (2007), "Gender Issues in Higher Education in Pakistan," *Bulletin*, Issue 162.

Kama, Lisapo ya. (2020), "The African Matriarchal Tradition," downloaded on March 31, 2020, from: http://en.lisapoyakama.org/the-african-matriachial-tradtition/.

Madagascar Express. (March 5, 2012), "Dear France, Thank You for the 2009 Coup," *Madagascar Express*.

Ministry of Higher Education. (July 16, 2016), *Report on the Review of Private Higher Education Institutions*. Ministry of Higher Education unpublished report.

Mint. (June 23, 2021), "India Has Seven Varsities among the World's 200 Best Young Universities," Live Mint, downloaded on June 30, 2021, from: https://www.livemint.com/education/news/india-has-seven-varsities-among-world-s-200-best-young-universities-11624449551128.html.

Moleah, A. (1993), *South African Colonialism, Apartheid, and African Dispossession*, Wilmington, DE, Disa Press.

No author. (2009), "I Love South Africa but I Hate My Government." Downloaded on July 23, 2019, from: https://iluvsa.blogspot.com/2009/11/no-retribution-for-mbeki.html.

Quintana-Muret, Lluis, et al. (2010), "Strong Maternal Khoisan Contributions to the South African Coloured Population: A Case of Gender-Biased Admixture," *The American Journal of Human Genetics*, vol. 86, pp. 611–620.

Raza, Ahsan (September 2, 2020), "No Pakistani University in Top 500 Times Higher Education World Rankings 2021," downloaded from: https://en.dailypakistan.com.pk/02-Sep-2000/no-pakistani=university-in-top-500-times-higher-education-world-rankings-2021.

Razafindrakoto, Gaby. (2016), "Gender Equity in Madagascar," *World Policy*, downloaded on September 17, 2019, from: https://worldpolicy.org/2016/03/15/gender-equity-in-Madagascar/.

Rotberg, Robert I. (2012), *Transformative Political Leadership*, Chicago, IL, University of Chicago Press.

Sen, Amartya. (1999), *Development as Freedom*, New York, Random House.

State University. (n.d., circa 2020), "Hybrid Learning Flexible Online Option," downloaded on June 26, 2021, from: https://education-state.university.com/pages/1146/Pakistan-HIGHER-EDUCATION.html.

Stevens, Siaka P. (1984), *Siaka Stevens: What Life Has Taught Me*, Bucks, England, Kensal Press.

Thomas, Roger. (1974), "Education in Northern Ghana, 1906–1940: A Study in Colonial Paradox," *International Journal of African Historical Studies*, vol. 7, no. 3, pp. 127–164.

Trading Economics. (n.d. circa 2021), "Madagascar Corruption Rank."

Ullah, Hazir. (September 30, 2020), "Quality in Higher Education: What Is Wrong?" *The Nation*, downloaded on June 26, 2021, from: https://nation.com.pk/Reporter/dr-hazir-ullah.

UNICEF Ghana. (n.d., circa 2020), p. 4.

UNICEF. (2020), "At a Glance: Sierra Leone Statistics," downloaded on May 7, 2020, from: https://www.univef.org/intobycountry/sierralene_statistics.html.

Wellcome Trust Sanger Institute. (2005), "Human Inhabitants of Madagascar are Genetically Unique," *ScienceDaily*, downloaded on September 16, 2019, from https://www.sciencedaily.com/releases/2005/06/05068102652/htm.

Wikipedia. (2021), "Women's Education in Pakistan," downloaded on June 25, 2021, from: https://en.wikipedia.org/wik/Women%27s_education_in_Pakistan.

World Bank. (2006), "Higher Education Policy Note Pakistan: An Assessment of the Medium-Term Development Framework," Human Development Sector, South Asia Region, Note 1.

World Bank. (2018), "Country Report," p. 4, downloaded on April 11, 2020, from: https://www-bti-project.org/en/reports/country-reports/detail/itc/sie/itr/wca/.

World Bank. (2020), "The World Bank in Sierra Leone," downloaded on April 11, 2020, from: https://www.worldbank.org/en/country/sierraleone/overview.

World Bank. (2021), "The World Bank in Sierra Leone," Washington, DC, Word Bank, pp. 1–3.

Yahoo News. (March 4, 2021), "12 Indian Institutes in top 100 of QS World University Ranking; Pokhriyal Credits Govt's Education Reforms," downloaded from: https://in.news.yahoo.com12-indian-institutes-top-100-16245124.html.

Index

F. M. Hayward, *Fostering Institutional Development and Vital Change in Africa and Asia*,
https://doi.org/10.1007/978-3-031-04364-2

FSC
www.fsc.org
MIX
Papier | Fördert
gute Waldnutzung
FSC® C083411

Kolibri 360 GmbH
Ferdinand-Jühlke-Straße 7
99095 Erfurt, Deutschland
produktsicherheit@kolibri360.de